Nothing to Lose E

'An incisive analysis of the impact of twenty-first-century capitalism on work that charts the creative ways in which workers are fighting back against modern day exploitation.'

—John McDonnell, Member of Parliament for Hayes and Harlington

'Shows the stark reality that, while we have developed more creative ways of winning and seem to be winning more, the impact of capitalism and exploitation of workers hasn't changed very much at all.'

—Sarah Woolley, General Secretary of the Bakers, Food and Allied Workers Union

'A much-needed look at one of the biggest issues for employment relations research and trade unions today: precarious workers. Any study of contemporary union organising that embraces rank and file militancy as a way of building networks of solidarity is a welcome contribution to the debate.'

—Dave Smith, co-author of *Blacklisted: The Secret War Between Big Business and Union Activists*

'Deserves to become a guidebook for labour movement activists that can help to further energise collective resilience and resistance.'

—Ralph Darlington, Emeritus Professor of Employment Relations, University of Salford

'We have a decision to make: we can sit back and hope the trade unionists of tomorrow will emerge, or we can fight together for the future the next generation deserves. *Nothing To Lose But Our Chains* inspires us with contemporary and ongoing tales of fighting and winning.'

—Rohan Kon, Organiser for Sheffield Needs A Pay Rise

'A welcome reassertion of the crucial inter-relationship of gender and class in the struggle between labour and capital, placing recent industrial action by women workers centre stage.'

—Sian Moore, Professor in Employment Relations and Human Resource Management, University of Greenwich

Nothing to Lose But Our Chains

Work and Resistance in Twenty-First-Century Britain

Jane Hardy

First published 2021 by Pluto Press
345 Archway Road, London N6 5AA

www.plutobooks.com

British Library Cataloguing in Publication Data
A catalogue record for this book is available from the British Library

ISBN 978 0 7453 4103 3 Hardback
ISBN 978 0 7453 4104 0 Paperback
ISBN 978 1 78680 810 3 PDF
ISBN 978 1 78680 811 0 EPUB
ISBN 978 1 78680 812 7 Kindle

This book is printed on paper suitable for recycling and made from fully managed and sustained forest sources. Logging, pulping and manufacturing processes are expected to conform to the environmental standards of the country of origin.

Typeset by Stanford DTP Services, Northampton, England

Simultaneously printed in the United Kingdom and United States of America

Contents

Figures

Tables

Acknowledgements

First and foremost, I am hugely grateful to the workers who gave up their time to be interviewed for this book. Many rank-and-file activists and trade union organisers generously shared their inspiring stories of struggle. I applaud Caroline Johnson and Mandy Buckley and the Birmingham women carers in Unison who so tenaciously fought to defend their jobs and wages. Thanks go to the Unison Branch in Glasgow, both to full-timers Jennifer McCarey and Mary Dawson and to activists such as Lyn-Marie O'Hara who fought for and who won such an important victory for equal pay. Thanks also to Rhea Wolfson at the GMB. Sandy Nicoll was hugely helpful, not only in furnishing me with the details of the long battle of his Unison branch in SOAS against outsourcing, but also in stressing the importance of politics and trade unions. In addition, I benefitted enormously from the interview with Henry Chango-Lopez and the story of the fight both with their employer and trade union at Senate House, University of London. I would like to thank Katie Leslie from the PCS for sharing her account of the strike by low-paid workers who punched way above their weight and the five cleaners who shared their enthusiasm for their struggle. Luke Primarolo and Cheryl Pidgeon from Unite were generous in sharing the details and challenges of the fight they waged against appalling conditions at the Sports Direct warehouse. I am in awe of the unswerving support provided by rank-and-file members of the Unite Community branch: Jeannie Robinson, James Eaden and Aubrey Evans in particular. Magda (not her real name) shared her story about her uphill battle as a union rep working and organising on the warehouse floor.

I have worked in education all my life. I started my working life in an inner-London comprehensive in 1977, worked in a further education college and then in higher education from 1992 onwards. I have always been a rep, on the branch committee or in the case of the UCU on the National Executive Committee from its inception in 2006 until six years ago. Thanks go to stalwarts of the UCU and its predecessors, in particular Liz Lawrence, Malcolm Povey and Tom Hickey, for helping me reflect and remember. Sean Wallis, Anne Alexander, Carlo Morelli, Mark Pendleton and Lesley McGorrigan shared their inspiring accounts and anecdotes of the recent struggles in universities over pensions, inequality, casualisation and pay. The achievements of teachers, organising under lockdown and forcing the government to back down twice on fully opening schools, has been amazing. I applaud the energy and ingenuity of teacher activists Chris Denson, Venda Premkumar and Emma Davis. Thanks go to Jon Hegerty for sharing his knowledge and expertise about the NEU, which set an example to the trade union movement about how to protect workers and communities during a pandemic.

Last but not least I would like to thank interviewees who could be considered the 'new kids on the block' – those workers and trade union organisers who have shown that there are no no-go areas for labour organisations. I am hugely grateful to Gareth Lane, Bryan Simpson, Bob Jeffreys, Austin Kelmore, Sarah Hughes, Max, Ava Caradonna and Jason Moyer-Lee who have organised in new sectors of the economy or those that are deemed to be unorganisable. Their struggles have struck a blow against those that argue that young people are not interested in trade unions.

The second group of people to whom I am indebted are those who took time out their busy schedules to read and comment on all or parts of the book. The contribution of Joseph Choonara, who read a complete early draft, is much appreciated. He offered important and insightful comments. Also, thanks go to Richard Milner and Bob Jeffreys – unknown to me before I undertook this project – but who have both been encouraging and generous in making useful suggestions. The content on trade union struggles has benefitted enormously from the forensic knowledge and political arguments, honed over a long period of time, of Dave Lyddon and Ralph Darlington. Discussions with Mark Thomas were extremely helpful and I am grateful for the insights provided by the work of Yuri Prasad on black struggles and Martin Upchurch on work and workers' organisation. Thanks to Simon Joyce for sharing his expertise on 'gig' workers and to Paul Stewart and Xanthe Whittaker for early comments on the book proposal.

I would like to acknowledge the joint research that I did with Nick Clark and Ian Fitzgerald on the response of trade unions to the arrival of workers from Central and Eastern Europe in the period after 2004. Also, I have greatly benefitted from the Polish workers – in Poland and Britain – who shared their experiences of being migrants. I appreciate the help of Maciek Bancarzewski and Julia Kubisa who acted as translators in capturing their stories. I was honoured to be the guest of ex-miner David Wray at the Durham Miners' Gala in 2019 and to witness the vibrancy and resilience of working-class culture by both older and new generations. Many thanks to Charlie Kimber for helping me source some fantastic photographs and to photographers Guy Smallman, Geoff Dexter, Andrew McGowan and John Sturrock for taking them. The team at Pluto have been extremely helpful. Particular thanks go to David Castle for his suggestions on the first draft, which helped me sharpen up the final version, and to Robert Webb for his patience in managing the production process, and also to Dan Harding for copyediting the entire manuscript.

Finally, family and friends have been an invaluable source of information, expertise and encouragement. Iain Hamilton has been patient and beyond helpful in sorting out technical issues with my computer and John Hill has been generous in sharing his rigorous knowledge of published statistics. Friends have stepped up to fill my deficits with grammar and syntax. Therefore, thanks go to Viv Bailey and Jon Berry for casting their eagle eyes over several chapters, and to Kate Hunter for the political acumen and professional expertise she brought to bear on the book. I am proud of my daughters Kate and Shan for their activism. Gratitude goes to Kate Hardy for her generosity with contacts, advice and her critical gaze. Also, thanks to Shan Hardy for her story about holding a prosecco and samosa party to get out the vote for a strike on pay in the East London school where she was a rep. Thanks to Keith Randle for helping me enrich the argument about the nature of knowledge and creative work through his research on media, film and pharmaceutical workers. Alan Fair has been a constant support and helped enliven my writing by exhorting me to use rhetoric more effectively. Of course, I fully appreciate that not all the people that I have interviewed or acknowledged will agree with my analysis. But I hope that my arguments will be viewed as a continuation of and contribution to lively, important and comradely debates about struggle in which lots of voices need to be heard.

Abbreviations

ANL	Anti-Nazi League (set up 1977)
APEX	The Association of Professional, Executive, Clerical and Computer Staff (in 1989 APEX merged with the GMB trade union and now exists as a section within it)AUT Association of University Teachers
BAME	Black, Asian and minority ethnic
BECTU	The Broadcasting, Entertainment, Communications and Theatre Union (became a sector of the Prospect trade union in the United Kingdom on 1 January 2017 following the merger of BECTU with Prospect)
BEIS	Department of Business, Energy and Industrial Strategy
BFAWU	The Bakers, Food and Allied Workers' Union
BTZ	Better than Zero
CSU	Civil Service Union (1917–88)
ESOL	English for speakers of other languages
FBU	Fire Brigades Union
GMB	General and Municipal Boilermakers Union, now simply known as GMB, is a general trade union. It has 631,000 members across industrial sectors (retail, security, schools, distribution, the utilities, social care, NHS, ambulance service and local government)
IWA	Indian Workers' Association
IWGB	Independent Workers' Union of Great Britain
IWW	Industrial Workers of the World (byname Wobblies, founded in Chicago in 1905 by representatives of 43 groups)
LGBT	lesbian, gay, bisexual and transgender
NATFHE	National Association of Teachers in Further and Higher Education
NEU	National Education Union
NFWW	National Federation of Women Workers (established 1906 as a general trade union open to all women across a range of industries where women's work predominated)

NHS	National Health Service
NUM	National Union of Mineworkers
NUT	National Union of Teachers
NUTGW	National Union of Tailors and Garment Workers
ONS	Office for National Statistics
PCS	Public and Commercial Services Union
PFI	Private Finance Initiative
PPE	personal protective equipment
REF	Research Excellence Framework
RMT	National Union of Rail, Maritime and Transport Workers (founded 1990)
SEIU	Service Employees International Union
SNP	Scottish National Party
SOAS	The School of Oriental and African StudiesTEF Teaching Excellence Framework
TELCO	The East London Communities Association
TGWU	Transport and General Workers' Union (merged with Amicus to become Unite in 2007)
TUC	Trades Union Congress
UCU	University and College Union (founded in 2006 from a merger of the NATFHE and the AUT
UNISON	The UK's largest union with 1.3 million members representing workers in public services
Unite	Unite the Union (in 2007 the Amicus trade union merged with the Transport Workers Union to form Unite the Union; it has over a million members and represents workers in Britain and Ireland across 19 sectors of the economy: public services, transport, food, finance and construction)
UPW	Union of Post Office Workers (founded in 1919, it eventually merged with the National Communications Union in 1995 to form the Communication Workers' Union)
USS	University Superannuation Scheme
UVW	United Voices of the World
ZHCs	zero-hours contracts

1
Changing Terrains of Work and Struggle

NEW LANDSCAPES OF WORK

In 1993 the aircraft manufacturer, British Aerospace, located in the small town of Hatfield (Hertfordshire) 20 miles from London, ceased production and the site was handed to developers. This was part of a massive restructuring of British Aerospace, which slashed its entire workforce by almost half from 127,000 to 60,000; 40,000 of these jobs were from its aircraft division. The empty shell of a workplace in Hatfield was a sharp contrast to the human tide of workers flooding into the factory for the 7.30 am shift that I had witnessed only the year before. Skilled engineers and office and maintenance workers arrived on foot, or by bike, bus or car. The aircraft factory, opened by the de Havilland company in 1930, had been an icon of British industry. It was at the forefront of aerospace technology with the development of the Comet, the world's first commercial jet airliner in the 1950s. In 1960, de Havilland was taken over by Hawker Siddeley, and the innovation of the Trident medium jet airliner was one of the most significant engineering achievements of the twentieth century. The contraction of the market for commercial aeroplanes, increased competition in the sector and financial problems contributed to the demise of the Hatfield site with the loss of 3,000 jobs.

With its sleek white art deco buildings, what eventually became British Aerospace when the industry was nationalised in 1978 had physically and economically dominated the local economy for more than five decades. With its production facilities, airstrip and hospitality suite for selling planes to customers from some dubious regimes, the site had occupied 400 square acres. The factory was at the hub of the town, with its vibrant social club and sports facilities fielding sports teams that were second to none. Such was its global reputation that the Hatfield Technical College (the forerunner of Hatfield Polytechnic and then, in 1992, the University of Hertfordshire) was established in the 1950s to train aerospace engineers from all over the world. The factory was highly unionised with a systematic structure of shop stewards and, as in all sections of British manufacturing, at least a handful were members of the Communist Party. I was told by a former employee that in the 1930s workers in the factory paid for materials and constructed stretchers to fit onto the sidecars of motorcycles for use by the Republicans in the Spanish Civil War.

Over the three decades following the closure of British Aerospace the redundant land and buildings were completely transformed. Nobody could miss the irony of the fact that in 1997 one of the first uses for the site that had developed the Mosquito (then the fastest plane in the world that was pivotal in combat during the Second World War) was as a location for shooting Steven Spielberg's war film, *Saving Private Ryan*, set in the same time period. The fictional 'ruined' French village of Ramelle was built on the abandoned site of the British Aerospace factory. The set was reused in 2001 for Steven Spielberg's television series *Band Of Brothers* and then disassembled. But the much-vaunted Hertfordshire film cluster failed to materialise as competing film locations in Europe offering more substantial subsidies became more attractive.

A large tranche of the redundant British Aerospace site was developed as a business park. As part of the mass expansion of higher education in 2003, the University of Hertfordshire closed two of its rural campuses and consolidated them onto one large 'state-of-the art' campus built under the Private Finance Initiative (PFI) on the new site. The aircraft hangar, a listed building, is now David Lloyd Hatfield, reputed to be the largest health club in Europe, and the old control tower now houses offices. One particularly elegant art deco building was converted into the Hatfield Police Station outside of which there was a Black Lives Matter demonstration in August 2020. The business park is dominated by huge logistics firms who are moving things (Booker, Yodel, Royal Mail) or providing the technology for moving things (Computacenter) or are warehouses for storing things (Ocado). The business park houses the headquarters of multinational companies, spawned from the privatisation of British utility companies, in shiny offices (EE phone company, Affinity Water).

Over the three decades since the closure of British Aerospace the workforce has changed profoundly. Houses that had been the homes of British Aerospace workers were bought up by landlords and rented on a multiple occupancy basis to students and the migrant workers who had started to arrive, particularly from Central and Eastern Europe, after 2004. Such was the growth of the Polish community that, in cooperation with the university and on its premises, in 2011 a visit and talk was hosted for Lech Wałęsa, hero of the Solidarity uprising in 1981, as part of his whirlwind and lucrative tour of Britain. The shift to the service sector has not meant the evisceration of workplace strife – both the university and the giant Post Office distribution centre have been sites of strikes and picket lines. In November 2011, as part of a national strike of 2.5 million public sector workers protesting against detrimental changes to pensions, thousands of workers in Hertfordshire marched and demonstrated locally, culminating in a rally of unprecedented size at County Hall in Hertford.

This story of economic change has brought about a metamorphosis in employment in the town. The university, with its 20,000 students, is the largest employer and a hub for promoting 'local development', while the giant logistics firms on the business park have opened up more and new types of jobs in warehousing, transport and delivery. This journey through highly skilled aircraft production to filmmaking, distribution and education is emblematic of the restructuring of the British economy. There is nothing unique about changes to the local economy and work in Hatfield; the towns, cities and regions of Britain have all undergone transformations. The only differences are the timescale and the details.

From the early 1990s, along with changes in the economic structure of Britain, a pattern of historically low levels of strikes emerged. In every year since 1991 the number of strikes has been lower than the number of strikes in any year prior to 1991, and by 2020 this trend had continued unabated. Structural changes in the British economy, epitomised by Hatfield, have been cited as a key factor undermining the collective potential of workers, as traditional areas of the economy have been replaced by innovative forms of production and changing ways of consuming. In the early 1980s the decline of manufacturing gathered pace and there has been an ascendancy of finance as, since 1979, the City of London has been courted and nurtured by successive governments. New categories of work have emerged, with some workers labelled as knowledge or creative workers who are, it is argued, qualitatively different and more difficult to organise into trade unions. The 'gig' economy has brought in its wake an explosion of casualisation in the form of short-term and zero-hours contracts (ZHCs), seen by some as a major explanation for the low level of trade union struggle.

One narrative is that work and the possibilities of resistance are different and more difficult under the pervasive influence of neoliberalism. A subtheme is that the financialisation of the economy, and interlinked trend of outsourcing, has pitted workers against each other and makes it more difficult for them to identify common interests. Specifically, it is claimed that divisions are entrenched between public and private sector workers and the young and old. With the end or decline of heavy industries such as iron and steel, coal mining and engineering, some point to the destruction of working-class communities and the solidarity that was embedded in them. A dominant narrative is that cooperation has been replaced by a fractured and individualistic labour force that is a barrier to collective action in the workplace. This is associated with pseudo-psychological explanations that focus on the neoliberal self, where the hegemonic idea of individuality has permeated everyday life to such an extent that young people are apathetic or even hostile to joining trade unions. Others have blamed the draconian anti-trade union legislation introduced by Margaret Thatcher in the 1980s, further tightened in 2016 by David Cameron's Conservative government, for strangling the ability of workers to take industrial action through the use of the legal apparatus of the state.

THEMES OF THE BOOK

Challenging Mainstream Narratives of British Capitalism

There have been many attempts to write off the working class. The classic text, *The Affluent Worker in the Class Structure* (Goldthorpe et al., 1969), suggested that collectivity had been replaced by complacency as workers were 'too well off', while *Farewell to the Working Class*

(Gorz, 1997) laments the disappearance of the skilled worker who is the agent of change and author of the 'socialist project'. More recently *The Precariat: The New Dangerous Class* (2011) by Guy Standing has reignited debates about the changing nature of the working class by suggesting that there is a sharp division and even mutual antagonism between those in stable and permanent work and a new group, the 'precariat', who experience unstable work and ZHCs. The first aim of this book is to challenge some of these grand narratives about capitalism in general, and British capitalism in particular, and the changes in work that it has brought about. Labels given to changes in capitalism since the early 1970s include post-Fordism and the post-industrial economy. These have been superseded by epitaphs of the knowledge economy and financial capitalism, and in the current period the 'gig' economy is imbued with totemic importance.

Chapter 2 discusses how narratives of neoliberalism have fed pessimism about the combativeness of workers. I argue that accounts of global capital mobility that leave workers powerless have been exaggerated. The importance of the financial sector is highlighted and, in particular, the interconnection between financialisation, privatisation and outsourcing is underlined – in the care and higher education sectors for example. However, workers have not been passive in the face of Britain's particularly financialised version of neoliberalism and later chapters are a testament to the successful struggles of workers who have been challenged, but far from disarmed, by outsourcing. Chapter 3 dissects two other stories of British capitalism. The argument of the 'weightless' economy that relegates producing real goods to the sidelines of economic activity is taken apart by arguing that the line between manufacturing and services is blurred and that the creation of surplus value is dispersed throughout a chain of activities. Claims that new innovations in technology are game changers in terms of the amount of employment and the nature of work itself with the rise of a new breed of 'knowledge' and 'creative' workers are put under scrutiny. I argue that these categories are arbitrary and artificial and that this hype is not only divisive, but ignores the underlying labour process. Chapter 4 engages with debates about the new icons of neoliberalism – namely the gig economy and associated insecure work. Widespread coverage of precarious work by the progressive media and academics has been important in shining a spotlight on some of the scandalous conditions in which some people work for unscrupulous bosses, but is in danger of inducing pessimism and a hand-wringing fatalism. In order to tease out what practices are new and what are hangovers from the past I argue that many contemporary discussions about precarious work lack rigour in terms of definitions, data and historical continuities. In these contemplations of the current economy, the history of work is woefully neglected. The reason for placing struggles in a longer historical sweep is not just to celebrate past victories of workers, but to inform current debate and tease out lessons for activists in terms of what are continuities and what are cleavages over time.

Taken together, the purpose of these three chapters is to recognise and document the dynamic change that capitalism brings to the structure of economies and transformations in the type and nature of work. An understanding of developments in global neoliberalism and how these are refracted in the particularity of British neoliberalism, in a more general sense and in the workplace, lays bare the new challenges for workers and organised labour. However, relabeling the institutional arrangements of capitalism, focusing on its superficial characteristics and proposing new categories of work that divide workers from one another, distracts from its underlying drivers. In proposing that there is an inherent conflict between capital and workers in capitalist production, with exploitation at its heart, Marx provided the tools for understanding that all workers have a common cause. The term 'exploitation' is more than a negative adjective or a judgement about where bosses lie on the Richter scale of meanness. Marx explained that workers are not paid fully for the hours of labour that they expend in making goods or producing a service – in other words they create surplus value, the source of all profit. But the existence of competition, inscribed in the DNA of the capitalist system, means that bosses are constantly forced to defend their profits through extensive exploitation (longer working hours), intensive methods (harder work) or lowering wages. Whether a worker is a cleaner, a warehouse operative, a games developer or a university lecturer, exploitation lies at the heart of capital–labour relationships. Therefore, the second theme of this book is an exploration of the institutional response of workers to exploitation – namely trade unions.

Working-Class Struggle and Trade Unions

Trade unions grew out of the collective response and struggles of working people to poor wages and harsh working conditions. Engels (1881) points out that because capitalists are always organised and have the full backing of the state, 'the work people from the very beginning cannot do without strong organisation, well-defined rules and delegating its authority to officers and committees'. In the first part of the nineteenth century trade unions were the preserve of conservative craft unions, but upheavals at the end of the century (in particular the Match Women's strike of 1888 and the Great Dock Strike of 1889) brought unskilled workers into trade unions in their tens of thousands – groups of workers who today would be labelled as precarious. Even in the current period establishing a trade union in a workplace and going on strike meets with harsh and aggressive responses from some employers. Witness Michael Ryan, CEO of Ryanair, who boasted that 'hell would freeze over' before he had unions in his company, being forced to back down, negotiate with striking pilots and recognise trade unions (Stern, 2017).

The barometer of workers' militancy is conventionally measured in working days lost through strikes (ONS, 2020a) that are taken as a rough guide to the ebb and flow of struggles over time. In Britain there has been a long-term downward trend; 2020 saw the lowest level of strikes since data collection began. A pessimistic reading of these dismal figures might give succour to governments and employers, who see this as a shift in the frontier of control that gives them carte blanche to hold down pay, introduce practices detrimental to working conditions and bully employees with impunity. However, although strikes are the highest and most visible form of struggle, resistance by workers takes many and varied forms. Throwing 'a spanner in the works' to stop the production line has a long history as a way of protesting against the boredom and intensity of work and is alive and kicking under contemporary capitalism. The contradiction in capitalism between workers – who generate wealth and have nothing to sell but their labour – and those that own the means of production embeds conflict in the workplace.

Accounts of contemporary workers' struggles are foregrounded in Chapters 5 and 6. These explore the central contradiction in the institutional form that this conflict takes – that is, trade unions. Mainstream accounts of industrial relations can be arid and superficial, treating trade unions as monolithic and uncontested organisations and thereby conflating the interests and aspirations of ordinary members and the leadership. But the dual nature of labour organisations in managing discontent pulls them in different directions as they try to walk the tightrope between currying favour with employers and meeting some of the aspirations of their members. C. Wright Mills (1948) referred to the ambivalence of trade union leaders in leading and supporting struggles by dubbing them 'manager(s) of discontent', where their role was to control and defuse explosions of anger. Chapter 5 sets the balance of power between the state and employers on the one hand and workers on the other, and within unions in the historical context of the post-war era, showing how this balance may veer between high and low points of workers' struggles, with explosions of militancy and bitter defeats. Chapter 6 sheds light on the experience of many activists who have had to wage battles *inside* their trade unions to secure support from the leadership and other echelons of the bureaucracy and gain financial resources and solidarity with which to fight against intransigent employers. It opens the 'black box' of trade unions to tease out the dynamics between the leadership, paid officials and rank-and-file activists.

A major caveat is that I do not claim to systematically cover 'industrial relations' in this period. Rather the focus is on case studies of important campaigns and strikes that are either completely absent from, or covered cursorily by, the mainstream and even left-wing media. This is both to shed light on the 'ingredients' of successful struggles and to give a voice to workers from below.

Women and Migrant Workers Centre Stage

Gone is the prevailing image in the two decades following the Second World War of workers and trade unionists as white, male, manual workers. Therefore, the third aim of the book is to examine the changing landscape of a working class in which women and migrant workers are now central. In 2020, although 50 years had elapsed since the Equal Pay Act of 1970 and 45 since the Sex Discrimination Act of 1975, the gender pay gap was 9 per cent for full-time employees, and the division of labour between men's and women's jobs remained stubbornly in place. Women are still having to struggle for the equality at work that they were promised five decades ago, both with employers and in getting their trade unions to fight for their legal entitlements. Women have been the sacrificial lambs of the Covid-19 pandemic. Affordable childcare has contracted, a disproportionate number of women have been furloughed and they have carried the lion's share of home schooling. There is a real threat that hard-fought gains for equality will go into reverse.

Chapter 7 looks at the struggles of women care workers in Birmingham and women council workers in Glasgow. The two-day strike in October 2018, by mainly women workers, in Glasgow and their victory in winning equal pay was hugely significant. They succeeded in bringing the city to a standstill and exposed the lack of resolution on the part of the top echelons of trade unions to resolve inequality in the workplace nearly five decades after equal pay and anti-discrimination legislation – yet this barely registered in the media. Also hidden from history has been the long dispute of women care workers, members of the Unison trade union, with Birmingham City Council. In May 2019, after nearly two years and 82 days of strike action over the imposition of new contracts that would have slashed their hours and wages, the council backed down and the women won a momentous victory.

Chapter 8 focuses on migrant workers on whom British capitalism has always relied to fill labour shortages, for example Irish workers in the construction of the railways in the nineteenth century. After 1945 the country drew on its fading imperial past by inviting workers from the Caribbean to work in public services such as transport and the National Health Service (NHS), and from the Indian subcontinent to provide labour for the textile industry in Yorkshire and Lancashire and the engineering factories of Birmingham. In 2004, acute shortages of labour motivated the Labour government to make Britain one of three countries to fully open its labour market after the accession of the (mainly) post-Communist countries of Europe into the European Union (EU). The labour force was profoundly changed as the number of workers who arrived was estimated to be between 600,000 and one million, the majority of whom were Polish. Despite the stream of xenophobia from successive Conservative governments, and more veiled anti-immigrant rhetoric from some sections of the Labour Party, the dependence of the British economy on foreign-born workers has increased, with their share of the labour market increasing from 7.2 per cent in 1993 to 17 per cent in 2017 (Migration Observatory, 2019). The reliance on foreign-born workers in some sectors is stark. But we live in uncertain times and with retail, manufacturing and hospitality the worst-hit sectors during the Covid-19 pandemic it has been estimated that 1.3 million foreign-born workers left the country in 2020 (O'Connor and Portes, 2021). Yet despite the concentration of migrant workers in low-paid and precarious work they have, as we shall see, been at the forefront of important struggles.

Microcosms of Struggle

The final aim of this book is to bring together the previous three themes by focusing on microcosms of struggle. Workplace grievances and action bubble up constantly and, more often than not, quickly dissipate; they are absent from the mainstream media and below the radar of official measures of industrial action. Ballots in favour of industrial action and a quick retreat by employers are not newsworthy. But when employers are intransigent, ballots that translate into strikes have the potential to change the workers who take part and to escalate into something bigger, involving more workers and raising political demands. If strike statistics paint a broad but incomplete picture, then studies of individual strikes are a microcosm of the interplay between the specific form of exploitation, the role of the state and the balance of power within trade unions themselves. However, the type and level of struggle cannot simply be read off economic circumstances and/or the existence of restrictive or permissive legislation. The anger and grievances of workers do not automatically translate into joining trade unions and/or taking industrial action. No mechanical formula exists to act as a conveyor belt, moving those who work for poverty wages in abysmal conditions into struggles for better pay and conditions. To understand why some industrial disputes take off and others do not, and why some action results in victory and others in defeat, we need to look at microcosms of struggle. The focus of Chapters 9 and 10 is to look at selected important disputes in the period 2015 to 2019.

Chapter 9 focuses on workplace battles waged by low-paid workers. In a fight against outsourcing, London cleaners – regarded as an organisationally weak group of mainly migrant workers – pitted themselves against powerful public sector institutions (universities and government departments) and multinational subcontracting firms and achieved significant gains in their wages and working conditions. This book is a celebration of successful struggles in a dismal period for trade unions, where the tenacity and resilience some groups of workers have shown in taking industrial action is humbling. But I also try to escape 'the fog of euphoria' (Hyman, 1980) by including one example in this chapter where the outcome of a campaign to organise has been disappointing. The case study of the Sports Direct warehouse is important because of its high profile as an example of poor work, extreme exploitation and surveillance in the employment of mainly migrant workers from Central and Eastern Europe. Despite being shamed in the media and the heroic efforts made by socialists in the locality and activists in the Unite trade union, the gains made have at best been modest because industrial action was never on the agenda.

Chapter 11 looks at how neoliberalism has proletarianised workers in education and brought them into confrontation with the government and their employers. In universities the blue paper that ignited accumulated grievances about heavy workloads, casualisation and stagnating pay was the seemingly arcane issue of proposed changes to pensions that resulted in 36 days of strikes between 2018 and 2020. The Covid-19 epidemic that spread globally in 2020 brought teachers into collision with the government who they forced to retreat twice on fully opening schools. In both cases these struggles generated new waves of rank-and-file activists and reinvigorated their local and national trade union structures.

Chapter 12 argues that a feature of the landscape of work in the new millennium is that there are now no 'no-go' areas for trade unions. This is especially important given claims by some that young people are not interested in joining trade unions and collective action. Yet young people have experienced the brunt of unemployment during the Covid-19 pandemic. By March 2021 almost two-thirds of people who lost their jobs were under 25. This chapter challenges that assumption. Beyond the drama of strikes and industrial action, the chapter includes a case study of games developers whose organising is at an embryonic stage. Although representing only a small group of workers, electronic games development represents the cutting edge of the 'new economy' and epitomises 'cool jobs'. The rhetoric and reality diverge sharply. I look at how these mainly young workers are subject to new forms of exploitation in the form of systematic overtime and rampant bullying, and how they are challenging them by organising in both established and newly formed militant trade unions. I discuss how other groups of workers, in hospitality and working as couriers, thought to be beyond the pale of trade unions, have organised successfully against precarious work. Finally, although still controversial in the labour movement, I focus on how support for the organisation and health and safety of sex workers has grown in mainstream trade unions.

These disputes enable an exploration of new terrains of struggle for labour organisations, where a series of mergers has swallowed up smaller trade unions to form mega organisations such as Unite, Unison and the GMB with between half and one and a half million members. At the other end of the spectrum there has been the emergence of very small and militant trade unions such as the Independent Workers of Great Britain (IWGB) and the United Voices of the World (UVW), founded in 2013 and 2014 respectively, that have advocated for and organised low-skilled and migrant workers as well as those in the gig economy, such as games developers and couriers. They have punched above their weight and have achieved some remarkable victories.

LIVED EXPERIENCES OF WORK AND STRUGGLE

Workers as Agents of Struggle

In the early post-war period sociologists, then a relatively new breed of academics, put workers under the microscope as *objects* of their dispassionate studies. In parallel, the study of industrial relations in the 1950s and 1960s was focused on instilling an orderly framework for trade unions to mediate between capital and the grievances of workers. This was in the face of what were perceived as unruly workplaces with frequent 'wildcat' strikes, where employers and the government claimed that shop stewards (workplace representatives) lacked accountability to the official trade union machinery. But in the late 1960s a handful of groundbreaking books recorded the lived experiences and testimonies of workers, giving them humanity and agency and making them the *subject* of the workplace. These seminal contributions include, for example, *Working for Ford* by Huw Beynon (1973) and *Girls, Wives, Factory Lives* by Anna Pollert (1981).

Accounts of strikes provide important lessons for activists. Between the late 1960s and mid-1970s there was an explosion of academic studies, films and plays that put the independent action of workers at their centre – reflecting the eruption of industrial struggles led by the rank and file. These sidelined the arid and neutral approach to industrial relations and were an antidote to anti-trade union films, such as *I'm Alright Jack* (1959) and *The Angry Silence* (1960), that dominated the 1950s and early 1960s. The first full-scale study was of the seven-week strike of 8,500 workers at Pilkington's glass factory in St Helens, Lancashire, in 1970, also the subject of the television play *The Rank and File* (1971). Both were important for exposing the strike's dynamics whereby disillusionment with the union (the General and Municipal Boilermakers' Union) prompted the workers to set up an independent strike committee (Lane and Roberts, 1971; Barker, 1970). Since then there have been other powerful accounts of individual struggles from the point of view of workers, relating to the explosion of strikes in the early 1970s by miners, dockers and building workers (Darlington and Lyddon, 2001); the Miners' Strike of 1984/5 (Callinicos and Simons, 1985); and the Liverpool dockers' strike in the mid-1990s (Lavalette and Kennedy, 1996). More recently, other books sympathetic to workers include coverage of the British Airways dispute (Taylor and Moore, 2019), the Grunwick and Gate Gourmet strikes involving Asian women (Anitha and Pearson, 2018) and the experience of working in a call centre (Woodcock, 2017). Socialists and academics have rescued other important disputes from falling into obscurity, for example the Lee Jeans occupation in Scotland (Robertson and Clark, 2019) and the Trico equal pay strike (Groves and Merritt, 2018), which are discussed in Chapter 7. The disputes covered in this book draw on this tradition of workers' self-organisation and their lived experience of struggle.

The Research

Case studies of struggles in the workplace provide the opportunity for a rich exploration of the dynamics of strikes and campaigns, each of which is unique. The workplaces covered in this book are set in the economic context of the given sector and how the specific conditions of exploitation are shaped by the product market for the goods or service and the nature of competition. I capture the lived experiences of workers – care workers, cleaners, games developers, warehouse workers, teachers and university workers – to give a flavour of labouring in the neoliberal era. The stories of struggles that are the focus of this book are told by the workers who took part in them. The case study disputes are drawn from the length of Britain – London, Birmingham, Nottingham, Sheffield and Glasgow – where I travelled to carry out face-to-face interviews between 2016 and 2019 and, from the end of January 2020, using Zoom. I have cited some interviewees, who were keen to be acknowledged by name; in other cases workers fearful of recriminations have been given pseudonyms.

Through interviews with rank-and-file workers, activists, elected branch officials and union organisers close to the workers, the narrative captures the accumulation of grievances and explosions of anger that are eventually provoked. I look at how events unfold, how strikers manage the day-to-day realities of keeping a dispute going in the face of intransigent and sometimes vicious managements, and how they fend off attempts by their own unions to foist negotiated deals on them that fall far short of their demands. Above all the case studies capture the role of a fresh layer of activists, the emergence of new leaders and the transformation of workers and their workplaces as a result of their participation in struggle. I am grateful for the time they found to speak to me and I am in awe of their enthusiasm, resilience and the resounding victories they have achieved, particularly at a time when the level of strikes is at a low ebb.

Finally, an important clarification is necessary in the discussion that follows. When I refer to socialists I am not talking about those who merely give themselves that label and sit on their hands, but about those who believe in democratic unions, the self-activity of workers and supporting action wherever it appears. As we shall see, they are not only catalysts for action in their own workplaces and committed to solidarity with other workers, but make links between 'bread-and-butter' fights and broader political struggles both at home and abroad.

NOT ALL IN IT TOGETHER

The hard-fought gains of workers – in defending wages and working conditions, in fighting for health and safety and against discrimination – are constantly threatened by capitalism. Driven by competition at the heart of the system, long-term structural changes in capitalism and innovations in technology and managerial practices have devastating effects on the lives and livelihoods of workers. But capitalism does not just change incrementally; endemic crisis means that it experiences sudden shocks, which have become deeper and more frequent since the global crisis and downturn of 1976. In the new millennium two convulsions have shaken the global economy to its core. In 2008 mountains of bad debt and the avariciousness of financial institutions resulted in a crisis in the very belly of the beast – the United States (US) – that spread throughout the globe and brought capitalism to its knees. Governments socialised the debts of the rich by pouring in eye-watering sums of money to rescue some of the largest and most profitable companies on earth in the banking sector and even the giant US multinational General Motors. To pull the entire system back from the brink they turned the debts of the rich into hardship for the working class. In Britain, as elsewhere, ordinary people were asked to shoulder the burden as ten years of austerity was imposed. Even by 2020 real wages have not recovered their pre-2008 levels, although the portfolios of the rich have been more than replenished. Deep cuts in the public sector have ravaged the welfare provision of national and local government and brought about new rounds of work intensification.

The second convulsion, in early 2020 and lasting into and possibly beyond 2021, was the Covid-19 pandemic that spread quickly from country to country and brought economies across the globe to a standstill as countries went into lockdown. The response of the British Conservative government and the prime minister, Boris Johnson, was bumbling, slow and desperate not to damage the profits of its allies in business. The rhetoric of being 'all in it together' was quickly exposed. The existence of vast and yawning inequalities and large pockets of insecure work were shown in sharp relief as thousands of workers lost their jobs in the first week, while other workers in warehouses, call centres and on building sites were coerced into going to work.

It became clear who were the key workers running society – health workers, teachers, checkout staff in supermarkets and people picking, packing and delivering food. Investment bankers were notable by their absence. Years of cuts to the budget of the NHS and the privatisation of its supply chain left front-line workers in hospitals and the care sector unprotected as necessary levels of personal protection equipment (PPE) failed to materialise. Yet the generosity of ordinary people shone through with support networks emerging spontaneously in communities to help the vulnerable: 405,000 people answered an appeal for volunteers by the government on behalf of the NHS on the first day.

In the immediate period, as I write, there are urgent questions as to how workers can collectivise through trade unions to defend their jobs, deliver high-quality goods and services and protect the health and security of their members. Even under conditions of a lockdown workers have had victories in making their workplaces safer as well as struggling for higher pay and against redundancies – battles that have gone under the radar of the media. But there are much bigger questions raised by the pandemic that go to the very heart of the organisation of the system we live in, which is how workers can contribute to building a sustainable system of production and a society that puts needs and life ahead of profits. This book makes a small contribution to that debate.

2
Neoliberal Britain

THE NEOLIBERAL ERA

There are few experiences that so dramatically lay bare the physical contrast of neoliberal Britain as the journey on the Docklands Light Railway from Bank station in the heart of the City of London. As you pass through the borough of Tower Hamlets the elevated track allows a clear view of one of the most economically and socially deprived areas of Britain. Its legacy of common-access 1950s public housing is punctuated by one council-built brutalist 1960s high-rise building and exorbitantly priced private apartments – often bought off plan for the investment portfolios of the global rich. Suddenly the financial district of Canary Wharf rises above its surroundings like a citadel – a protected fortress of global capitalism – with its centrepiece the shimmering Pelli Tower sheathed in stainless steel (see Figure 2.1). In 1981 Margaret Thatcher created the Docklands Development Corporation in East London, in which Canary Wharf is situated, where land was available free from planning controls and commercial rates with capital investment written off against tax. It was created as part of the neoliberal project and is a physical manifestation of it and of the ascendant role of finance in particular.

The borough of Tower Hamlets is a tale of two cities where the highest levels of child poverty in the capital exist alongside some of the most well-remunerated people in the world in the headquarters of financial institutions and banks (HSBC, Barclays, Morgan Stanley and J.P. Morgan) which recycle the spoils of global capitalism. The industrial legacy of the East End has been converted into luxury bijou flats – including the Bryant and May factory site of the Match Women's strike of 1888 – often bought as investments and left empty by absentee landlords. Only 10 per cent of the workforce of Canary Wharf live in Tower Hamlets: those that work there travel in and out

Figure 2.1 Council housing in Tower Hamlets with Canary Wharf financial centre in the background

Source: Abbie Trayler-Smith.

in hermetically sealed bubbles with, at best, token contact with the local community through worthy actions that salve their consciences and read well on their corporate social responsibility statements. At the other end of the labour market is an army of security guards and invisible workers cleaning the detritus of offices at dawn and dusk for a tiny fraction of the salaries of their occupants.

From the 1980s onwards profound changes in the economic structure and labour market of contemporary Britain are associated with the 'neoliberal' period. For activists and the left, neoliberalism is a shorthand pejorative term for universal and toxic developments in capitalism that led to markets being unleashed to permeate every aspect of work, welfare and personal life. It is a theory of economic-political practices which advocates that the well-being of humans can be brought about by putting in place strong property rights that free up entrepreneurship, liberate markets and reduce the role of the state. This new era of global neoliberalism was characterised by more porous boundaries for finance and capital, both reflecting and producing a further intensification of competition as barriers to trade and investment tumbled in developed and developing countries alike.

To understand changes in the types of jobs on offer and the nature of work itself from the mid-1970s onwards, it is important to consider the way that neoliberalism intertwines at the levels of the global, the national, the local and the workplace.

Global Neoliberalism and Mobile Capital

The first consolidation of the notion of neoliberalism as a coherent project can be traced back to the Mont Pelerin Society in 1947. A group of economists, whose leading lights included Milton Friedman and Friedrich von Hayek, drafted a statement advocating a limited role for the state and the centrality of private property and competition to steer reconstructing economies away from what they perceived to be the dangers of socialism.[1] But unprecedented rates of growth in the post-war period meant that neoliberal ideas had little resonance with economists and governments of both left and right. However, by the mid-1970s falling global growth rates compounded by the oil crisis and the weakness of the British economy sounded the death knell for, and exhaustion of, a set of arrangements by the ruling class often attributed to Keynesian economic policy. In 1976 the ensuing sterling crisis led to the Labour government going cap in hand to the International Monetary Fund, which demanded savage cuts in public spending. This was the beginning of neoliberalism in Britain that was to become an ideological crusade for the prime minister, Margaret Thatcher, after her election in 1979. The policies she implemented amounted to a project to reorganise British capitalism through the deregulation of finance and labour markets. Along with opening up a tranche of previously nationalised industries to privatisation she intended to decisively shift the balance of power away from (organised) labour.

Political upheavals in Central and Eastern Europe from the late 1980s and market-based reforms in China in 1992 enmeshed these countries more tightly with the global economy. Another fillip for neoliberalism was the introduction in 1993 of the Single European Market in the EU that jettisoned barriers to the movement of capital and labour. In so doing it set the stage for a frenzy of restructuring through acquisitions and mergers as capital moved freely across borders and consolidated the EU as a major economic and political capitalist bloc. New markets opened up and fresh sources of skilled and unskilled workers, often working at cheaper rates, were on tap through the free movement of labour.

With its incessant revolutions in the economic structure, Marx points to the contradictory consequences of capitalism as a source of misery and, at the same time, the source of all progress. Under the whip of competition that drives firms to adopt new innovations in technology, capitalism constantly reorganises itself within and between countries across the globe. Changes in the types of work on offer, and the nature of work itself and the new opportunities and challenges they bring for workers, have to be understood in the context of constantly changing patterns of capital and labour migration, and more specifically in the case of the British economy its integration with global capitalism.

However, it is important to put the dynamic changes of capitalism in perspective. For example, David Harvey (2005) argues that the ability of capital to move imposes increasingly precarious forms of work on workers, while Michael Hardt and Antonio Negri's *Empire* claims that 'Capital can withdraw from negotiation with a given local population by moving its site to another point in the global network' (2000: 297). The danger with these arguments is that they promote a view of workers as powerless and passive in the face of footloose capital. I have argued elsewhere that accounts of the wholesale mobility of capital are exaggerated (Hardy, 2013). While some sections of capital have always been relatively mobile – textiles and to a lesser extent electronics for example – the large sums required for investments in other industries, such as the auto sector or aircraft production, slow down their ability to move. Other parts of production, employing a large proportion of the labour force, are completely immobile and rooted in national territories, including the physical infrastructure (railways, airports, power plants) and the social infrastructure (education, health, welfare).

The dynamics and geography of capitalism to which Hardt and Negri and Harvey point can be read differently. In advanced capitalist economies the restructuring and relocation of manufacturing has taken contrasting forms and followed different rhythms. While it may be the case that labour has been weakened in the locations which productive capital has abandoned, new working classes have been created and strengthened in favoured sites of new investments, evident outside the core capitalist economies, for example in countries such as Brazil, South Korea and China (Silver, 2003). However, this dynamic of capitalism that destroys jobs in one place and creates others in new spaces has resonance in Britain. As mines and steelworks have closed in some parts of the country, new 'satanic mills' in the form of giant warehouses and call centres employing thousands of workers have opened up in other locations. In some cases, new forms of productive activities have replaced old ones in exactly the same physical space; for example, the warehouse of Sports Direct in Derbyshire, with its 4,000 workers, was built on the site of the Shirebrook coal mine that closed in 1993.

In practice neoliberalism is always and everywhere pragmatic – it is not a matter of imposing some blueprint with the clear endpoint of a pure free-market utopia. The logic of a completely free market would mean the reintroduction of child labour and turnstiles on every stretch of road. Neoliberalism has never meant complete liberalisation and/or the retreat of the state – rather it has been about the capturing and restructuring of the state in the interests of capital and the restoration of profit. The state and different sections of capital have different and sometimes divergent interests and therefore neoliberalism does not operate as a coherent ruling-class strategy – it is riddled with inconsistencies, obstacles and contradictions. For example, to compete with other capitalists some sectors rely on a stable workforce and their embedded skills. Such aspirations are incompatible with insecure work and a high turnover because it is expensive to lose workers and have to train new ones.

British neoliberalism went head-to-head with the collective neoliberalism of the EU in the form of Brexit. The EU is a power bloc predicated on the free market for labour and capital buttressed by a disciplinary mechanism to rein in public spending. In 2016 sections of the British ruling class won a vote for a populist and nationalistic version of neoliberalism arguing that potential global markets could compensate for diminished trade with the EU. Placing primacy on the market through deregulation, privatisation and labour flexibility is a common impetus of global neoliberalism, but how this process plays out in different nation states is highly variable and in Britain it has taken a very specific form.

THE PARTICULARITY OF BRITISH NEOLIBERALISM

The particularity of neoliberal Britain has three hallmarks. The first is its pivotal role in recycling global surplus value and the importance of the finance sector. The second is the way that the tentacles of privatisation have stretched into nearly every nook and cranny of the welfare state. Third, the interrelated role of outsourcing the tasks of local and national government have proved to be a source of easy and lucrative profits for large transnational corporations and a mechanism for attacking the wages and working conditions of workers.

Recycling Global Surplus Value

The pride of place held by the finance sector in British capitalism is revealed by the successive fawning and hubristic speeches given to bankers in the City of London by Labour's chancellor of the exchequer, Gordon Brown, and the Conservative prime minister, David Cameron. Specialised in providing financial services to the world, and second only to the US as an exporter of commercial services (including banking, asset management, trade in derivatives and insurance), Britain accounted for 6.5 per cent of world trade in the sector in 2018 (World Trade Organization, 2019). The City of London's pivotal role in international finance and commerce originated in Britain's grip on world trade in the eighteenth and nineteenth centuries and the rise of British imperialism, but its international importance has long outlived the decline of the British Empire. In 2019 the financial sector accounted for 6.9 per cent of total output in Britain and 3.2 per cent of jobs: geographically it is skewed towards London which accounted for nearly half of the output generated (Hutton and Shalchi, 2021). While the finance sector conjures up images of champagne-swilling, mainly white men, and ostentatious consumption on the backs of massive bonuses, the vast bulk of employment in the sector is in routine work, including badly paid call centres spread throughout Britain, with a culture of control and surveillance.

The ascendant and pivotal role of the City of London was not an accident of economics but the result of it being nurtured and promoted by successive governments. To try to retain Britain's competitive position in financial services, the liberalisation of the sector was driven by Conservative prime minister, Margaret Thatcher, beginning with the abolition of capital controls 1979. A set of reforms introduced in 1986 – collectively known as the Big Bang – swept away what were seen as traditional and archaic restrictive practices constraining the operation of the finance sector in the global economy. The New Labour government, from 1997 onwards, continued this love affair with the City with their 'light touch' approach and emphasis on 'self-regulation'. This put even greater distance between the activities of financial institutions and their democratic control and led to a huge rise of international finance capital in and out of the City of London. A dark underbelly of Britain's financial activities lies in its role in 'whitewashing' the money of kleptocrats and oligarchs from the ruling classes of other countries, that according to a Home Affairs Select Committee (2016) report 'laid out a welcome mat for launderers and organised criminals'. There have been dire prognostications by other members of the ruling class about the future of the finance sector after Brexit. It seems likely that there will be some fine tuning in global divisions of labour in finance; however, it remains to be seen whether Britain will lose its competitive advantage in the sector.

Finance, Finance Everywhere

Broadly, financialisation can be understood as referring to the increasing importance of finance, financial markets and financial institutions to the workings of the economy. While I argue elsewhere that these developments do not, as some propose, constitute a new stage of capitalism, it is the case that financialisation is a feature of and facilitated by the neoliberal era.[2] From the perspective of workers the intensification of financially driven targets in the context of sluggish profits has led to the squeezing of labour costs and the intensification of work. Grady and Simms go even further and present an apocalyptic vision of the impacts of financialisation on organised labour, arguing that it 'not only pits workers against each other, it pulls workers in multiple directions and makes it more difficult to identify a single set of interests' (2019: 497). But the victories of workers discussed in later chapters undermine this view of financialisation as a juggernaut that steamrollers workers' organisation and confidence. Nevertheless, partly as a result of the privatisation and outsourcing policies pursued by the governments of Margaret Thatcher and Tony Blair, finance has seeped into all aspects of the economy and everyday life. In the case of the New Labour government elected in 1997, their continuation of privatisation and outsourcing handed over great swathes of UK industry, public services and debt directly into the hands of London's financial sector and has enabled the legitimate and systematic raiding with impunity of the welfare state by the corporate sector.

The financialisation of care homes and universities has had profoundly deleterious consequences for the opportunities and welfare of working-class people, and the wages and working conditions of those employed in them. In 2020 the disastrous effects of the Covid-19 pandemic in care homes exposed the heinous state of a system opened up to the vagaries of financial markets after the Conservative government imposed compulsory tendering in the 1980s. This sector was seen as rich pickings for predatory international finance in the form of global private equity, sovereign and wealth funds attracted by the easy money to be made from a steady stream of government income and an ageing population. Corporate giants have used offshore tax havens and murky corporate structures to profit from the most vulnerable citizens in society. The stripping out of property from the balance books of care homes,[3] coupled with the introduction of the National Living Wage in 2016 and the impact of austerity as local authority budgets were slashed, squeezed the profit margins of these firms. This drove a trend towards bigger providers and behemoths, such as Four Seasons, HC-One, Bupa and Barchester, which used economies of scale to drive down costs – of which the biggest is staffing. Poor conditions of work and low wages were reflected in staff turnover that was among the highest in Britain, with almost a third of directly employed workers in social care leaving their jobs in 2019 (Skills for Care, 2019). This led, in turn, to a reliance on agency workers, often on ZHCs, who unwittingly spread the Covid-19 virus.

A major consequence of the nearly complete privatisation of universities through the replacement of government funding with income from student fees is that their property and the loans of their students have become coveted assets for global investors. The value of these outstanding loans at the end of March 2018 stood at £105 billion with a forecast that they will rise to £450 billion by the middle of the century (McGettigan, 2019). The government has sold on (or securitised) some of these debts at a significant loss for them to be snapped up by pension funds, insurance and investment companies at bargain basement prices (ibid.). Higher education is the new frontier for the bond market as universities raise money for large capital projects, with the market growing tenfold from £272 million in a single deal in 2007 to £2.4 billion by 2017 (Katsomitros, 2018). Building and renting university student accommodation have become rich pickings for international property investors. According to an article in *The Guardian* (Osborne and Barr, 2018) more than 20,000 students are paying for rooms owned by companies that have siphoned off profits to tax havens enabling them to avoid both income tax and stamp duty. As Chapter 10 explores, this financialised culture is both the outcome of and is driving competition in a higher education sector where huge, 'prestigious' capital projects lie at one end of the spectrum and casualisation and a harsh performance-related work culture lie at the other. The way that financial considerations trump educational ones in universities was sharply exposed during the Covid-19 pandemic in September 2020. In order to protect their bottom line, university managements were hell bent on getting students onto campus – in order to extract high fees and payment for expensive accommodation – even though most teaching was online.

Easy Pickings for Private Capital: Outsourcing

In line with the mantra of reducing the role of the state, outsourcing was a key component in the neoliberal agenda of Prime Minister Margaret Thatcher's government and has gone hand in hand with financialisation. Private sector involvement, enabled by successive legislation in the 1980s and 1990s, coerced competitive tendering in the provision of core council services and front-line public services such as prisons. Under the intensified rhetoric of 'choice and competition' Tony Blair's New Labour government enthusiastically pursued outsourcing, more than doubling public procurement in a decade and extending the role of private providers through, for example, the academisation of schools (Sasse et al., 2019).

New Labour greatly expanded the use of the PFIs – complex long-term financing agreements for public sector projects – that increased from a handful of contracts before 1997 to an average of 55 a year (ibid.). These schemes tied schools and hospitals into long-term 'facilities management' contracts where they were obliged to purchase porterage, cleaning and laundry, shackling them to costly deals that stretched for decades. Nothing was out of bounds as the Conservative-Liberal Democrat coalition government (2010 to 2015) extended outsourcing, prising open new markets in areas such as the probation service, and increased the use of private providers in health care and welfare. In 2018, £292 billion, more than a third of all public spending, went on purchasing goods, works and services from external suppliers by the government – up from £112 billion in 1987 (ibid.). Forty years on, transnational firms operate prisons, deliver probation services, assess benefits claimants, provide employment training, offer NHS operations and deliver social care.

A National Audit Office Report claimed that there was 'a lack of data available on the benefits of private finance procurement' (2018: 7), but what is certain is that these PFI schemes were a bonanza for private equity and transnational servicing firms, particularly in the early days when these firms 'made out like bandits' (Sasse et al., 2019). There is a plethora of reports, academic papers and publications documenting the catalogue of disasters and debacles from outsourcing. A string of contract failures that spanned security at the 2012 Olympic Games, prisons and young offenders' institutions and work capability assessments culminated in the spectacular collapse of the subcontractor Carillion in January 2018. This was one of the biggest corporate failures in the history of British capitalism and debts of £1.5 billion had to be picked up by the public purse. It had 420 contracts and employed 11,638 workers in Britain: they operated 218 school meals services, 50 prisons and 11,500 hospital beds, to give but a few examples of their vast empire. A parliamentary report in 2018[4] describes Carillion's rise and spectacular crash as a story of 'recklessness, hubris and greed' with a business model that was a relentless dash for cash, driven by acquisitions, rising debt, expansion into new markets and exploitation of suppliers (see Figure 2.2).

No lessons were learned from these scandals and the Covid-19 pandemic in 2020 proved to be another windfall for transnational private firms: often using personal contacts in government, and without having to go through a process of competition, they made a

Figure 2.2 Strike by NHS workers in Unison and Unite in Wigan that forces NHS trust to scrap plans to outsource 900 workers in June 2018

Source: *Socialist Worker*.

power grab for the plethora of contracts generated. Deloitte, KPMG, Serco, Sodexo, Mitie, Boots and American data-mining group Palantir secured taxpayer-funded commissions to manage the Covid-19 drive-in testing centres, the purchase of PPE and the purpose-built Nightingale hospitals. The privatisation of health and social care directly compromised the safety of health workers on the front line as the complex, inefficient and heavily privatised NHS supply chain proved to be a bureaucratic nightmare in getting PPE to front-line workers. Although nominally part of the NHS supply chain, in fact this was a fig leaf for a web of contracts operated by private companies answering to their shareholders.

Private providers have an incentive to reduce costs to maximise returns and profits for shareholders and the biggest potential for this is in the wages and working conditions of employees. The impact of outsourcing rears its head in and frames several of the case studies in this book. In Chapter 7 we see the way in which Glasgow City Council tried to circumvent equal pay by creating an arm's length company and how Birmingham City Council outsourced 'difficult' decisions to consultants. The struggle of cleaners in Chapter 9 looks at how contracts were bounced from one company to another, resulting in workers being on numerous different contracts. But although outsourcing was at the root of unequal terms and conditions and apparently complex organisational arrangement this was not, as we shall see, a barrier to workers organising to win stunning victories (see Figure 2.2).

NEOLIBERALISM IN THE WORKPLACE

Neoliberal ideas about the individual have their lineage in Adam Smith's *The Wealth of Nations* (2008), whose notion of self-interested workers and consumers underpins the 'invisible hand' of the market. In the new millennium a different idea of the individual under neoliberal capitalism has emerged and gathered pace in academic circles and among some activists. The idea of 'the neoliberal self' is based on Foucault's (2008) argument that there has been a linear progression from Adam Smith's self-interested 'economic man' centred on exchange to competitive individuals under neoliberal capitalism who are 'entrepreneurs of themselves' destined to constantly 'improve their human capital'. This has spawned a plethora of academic and popular articles and books that have analysed the penetration of neoliberal thinking in popular culture, education and work (Brown, 2015; Davies, 2016 and 2017). Foucault argues that the extension of the logic of the market into non-economic spheres of everyday life makes this reworked version of the individual all-encompassing and ubiquitous. There is no doubt that the neoliberal period has been characterised by a wide range of new managerial practices based on individual performance that include targets, bonuses and performance-related pay.

Resilience, the 'Wellness Agenda' and the Intensification of Exploitation

Employers have contributed to this enhanced version of the neoliberal self, for example by the introduction of a 'wellness agenda' with lunchtime nature walks and 'mindfulness' sessions. This allows employers to cultivate their image as benign and caring. But, more perniciously, it allows them to shift the blame on to individuals for their physical or mental health issues because they have failed 'to get with the programme', thereby deflecting attention from the intensification of work that has ratcheted up levels of stress in the workplace.

Precarious work is turned from a liability into employment that bestows the benefits of choice, flexibility and autonomy, a boundary-less career whereby the entrepreneurial worker packages her/himself and puts their portfolio on the employment market.

The idea of resilience has recently appeared in the lexicon of bosses, their acolytes and institutions. This is particularly pronounced in the NHS where long and intense working hours, heavy workloads and being on the front line puts doctors, nurses, paramedics and ancillary workers under extreme stress. But rather than address the chronic underfunding and unacceptable working conditions that produce this stress, healthcare staff are exhorted to be stoic. For example, an article that appeared in the *British Journal of Hospital Medicine* (Stacey, 2018) was entitled 'How to be a resilient doctor: skills to maximise your anti-fragility'.

Wellness has become big business. It fits like a glove with the ideological agenda of neoliberalism, making individuals responsible for their own health and happiness and diverting attention away from vast inequalities in physical and mental health under capitalism. Both 'wellness' and 'resilience' have been harnessed by governments as a potential way of raising the rate of exploitation. The British government jumped on the potential of 'mindfulness' to dismantle the mental and emotional obstacles to better performance and efficiency. The Mindfulness Initiative published a parliamentary report in 2015 that was transparent about how this agenda could be harnessed to get workers to take less time off sick with stress. It argues that developing 'cognitive and emotional resources that ensure resilience in the face of stress … should be of real interest to policy makers given the importance of improving productivity' (Mindfulness Initiative: 6). Not surprisingly it is bosses who have shown the most intense interest in and widespread experimentation with mindfulness. The report goes on to note that:

> What is driving the interest in and innovation of mindfulness in the workplace is the need to tackle issues around the *rising costs of workplace absence and absenteeism* because of the stress and depression and the *need to boost productivity* in a workplace which is being radically changed by new information technology. (Ibid.: 40, my emphasis)

Mindfulness, and other potentially therapeutic practices that offer some temporary respite from the stress and alienation of everyday work under capitalism, have therefore been hijacked and distorted as a tool for turning the blame on to individuals for their physical or mental illness and added to the arsenal of managers' weapons designed to get people to work harder.

However, although radical accounts of the neoliberal self make interesting and useful observations, they see these ideas as totalising and overestimate their coherence and the grip that they have on the minds of ordinary workers. Particularly problematic is their uncritical borrowing of terms from bourgeois economics that recast workers as 'entrepreneurs' possessing 'human capital' rather than possessing only their labour power: the use of this terminology eviscerates the central conflict between workers and capital in the process of production. Labour disappears as a category and so does the alienation and exploitation associated with work.

CONCLUSION

Accounts of neoliberalism that focus on the hyper-mobility of capital in global capitalism render workers as powerless and passive, while those that emphasise financialisation and outsourcing have drawn the conclusion that workers are fragmented and their collective organisations declawed. These gloomy views are summarised by Bourdieu who argues that the neoliberal project is a 'programme of the methodological destruction of collectivity undermining solidarity and encouraging individualisation' (1998: 95–6). The idea of fragmented and atomistic individuals at the heart of ruling-class ideology has echoes on the left and feeds into a general pessimism about the potential for the collectivity and solidarity of the working class and its ability to organise against its exploitation and to change society. But ruling-class ideas about the individual are not static; they are continually renewed, remoulded and defended and can therefore be challenged.

While the idea of individualism and self-gain may be embedded in the consciousness of highly paid steroid-driven workers in the financial institutions of the City of London coveting their bonuses, it is demonstrably not the case with workers in both the private and public sectors. Whether opposing tests and league tables in schools, performance-related pay or the outsourcing of cleaning, trade union activists have consistently struggled against neoliberalism in the workplace, as we shall see in later chapters. Further, the case studies show that solidarity between workers is very much alive. Chapter 3 explores how other narratives about the demise of manufacturing and the advent of new technology contribute to a pessimistic picture of workers rendered powerless in the face of unassailable capital.

3

Narratives and Numbers of British Capitalism

THE STORIES THEY TELL

On the 15 October 1997, newly elected Labour prime minister, Tony Blair, threw a party in Downing Street: guests included pop stars, comedians and actors, fashion designers and millionaires. The attempt to relaunch British capitalism as enterprising and creative was dubbed 'cool Britannia', but as we shall see this boosterism was window dressing for the enthusiastic continuation of the neoliberal policies of Margaret Thatcher. Neither was there anything new about this hyperbole. From the 1970s onwards there have been successive attempts to relabel British capitalism, spawning a variety of claims about the types of jobs on offer and the nature of work itself. These have ranged from wishful thinking about 'Britain PLC', and its branding and rebranding, to more grandiose claims that we are living in an entirely new phase of post-industrial capitalism. Heralding the end of mass production, and the arrival of flexible employment and apparently more satisfying work, by the end of the 1970s the narrative of post-Fordism[1] had taken hold, and in 2000 Hardt and Negri suggested that we live in a 'post-industrial economy, an informational economy' (2000: 285).[2] In Britain the advent of the 'weightless economy' was proclaimed by the Work Foundation where 'creating value depends less and less on physical mass, and more and more on intangibles such as human intelligence, creativity and even personal warmth' (Coyle and Quah, 2002). This story was extended by Will Hutton (2004), who argued that mass production in Britain was dead, with manufacturing moving to China and other parts of South East Asia while Britain concentrates on goods that 'are tailored, customised, niche products, exploiting research; in other words knowledge goods'.

In 2015 Paul Mason made even greater epoch-making claims for a post-capitalist world where technology is allowing some production to transcend capitalism itself in the 'niches and hollows of the market system' (2015: xv), where collaborative production using network technology 'defines the route beyond the market system' (ibid.: xvi). Even before this the impact of technology was a major preoccupation of politicians and academics. As early as 1964 the Labour prime minister, Harold Wilson, declared an ambition to 'replace the cloth cap [with the] white laboratory coat as a symbol of British labour' (1964: 54). Current iterations about the decline of material production and the impact of new innovations in technology assume the terminal decline of traditional manufacturing jobs and the replacement of boring work with flexible and fulfilling jobs based on creativity and new technology.

By distilling these arguments into two threads, this chapter dissects grand visions that claim tectonic changes in work and employment. The first story renders manufacturing as either peripheral or completely diminished, superseded by a weightless economy where the production of intangible or immaterial goods prevails. This is put under the microscope by looking beyond the numbers in order to argue that the boundary between manufacturing and services is far from clear and that the creation of surplus value is dispersed. The second story is one of unbridled economic determinism where changes in technology are claimed to be historically unprecedented and profound enough to constitute a 'fourth industrial revolution', where knowledge, creativity and emotion have replaced physical labour. It is argued that these, often dramatic, narratives make potentially divisive distinctions between different groups of workers and obscure the labour process.

A WEIGHTLESS ECONOMY? PUTTING MANUFACTURING IN PERSPECTIVE

There is a prevailing common-sense view that Britain has lost its factories and its manufacturing. The real story is much more complex and disentangling the myriad statistics is such a Herculean task that it is worth stating the headline propositions up front. I argue, first, that measuring 'manufacturing' in modern capitalism is not straightforward because the demarcation between producing goods and services is blurred; second, that Britain's slippage down the league table of economies in terms of manufacturing competitiveness has been long in the making; and, third, that real production still matters for British capitalism. Turning from grand narratives to numbers, Table 3.1 gives a thumbnail sketch of employment in Britain in the period 1997 to 2020.

Table 3.1 The composition of employment (selected sectors and years), 1997–2020

Year	*1997*	*2009*	*2020*
All employment	26,245	29,276	33,112
Public	6,075 (23.4)	7,224 (24.8)	7,239 (21.9)
Private	19,956 (76.6)	21,913 (75.2)	25,756 (78.1)
Manufacturing	4,370 (16.7)	2,856 (9.8)	2,981 (9.0)
Construction	1,891 (7.2)	2,482 (8.5)	2,319 (7.0)
Accommodation and food services	1,275 (4.9)	1,323 (4.5)	1,793 (5.4)
Information and communication	742 (2.8)	1,275 (4.1)	1,537 (4.6)
Finance	1,164 (4.4)	1,226 (4.2)	1,264 (3.8)
Professional, scientific and technical activities	1,497 (5.7)	1,841 (6.3)	2,651 (8.0)
Education	2,143 (8.2)	2,919 (10.0)	3,491 (10.5)
Human health/social work activities	2,869 (10.9)	3,771 (12.9)	4,529 (13.7)

Note: Absolute numbers in 000s; share of total employment as a percentage in parentheses.

Source: Author's adaptation from ONS (2020b): table EMP 13 Labour Force Survey.

The overall trend in employment is somewhat unexpected in that there has been a period of uninterrupted growth in jobs from 1997 onwards in nearly every sector, and even a 15 per cent increase after the crisis of 2008 and the austerity that followed. By 2018 the employment rate was 75.7 per cent – the highest since records began in 1971 (Clarke and Cominetti, 2019). In short, it tells a tale of an expanding job market with an overall increase in employment of more than 24 per cent, but with the private sector accounting for most of the growth in jobs.

Within this general picture of an increase in employment the exception is manufacturing, with a third fewer people working in the sector in 2020 compared with two decades earlier (with a brief resurgence of jobs between 2014 and 2018) and a contraction of its share in total employment. But in parallel there has been a massive expansion in jobs in education and welfare – employment that reproduces and maintains the labour necessary for human infrastructure of contemporary capitalism. Health and social work sectors accounted for nearly a quarter of the net increase in employment since 2008 (ibid.). For example, in 2019 the NHS employed 1.5 million people (about 1.1 full-time equivalents): it was the world's fifth-largest employer – exceeded only by the US Department of Defence, the People's Liberation Army of China, Walmart and McDonald's (Nuffield Trust, 2019). However, this table gives us only the very broadest of brush strokes; new forms of work are shoehorned into old categories – what are called standard industrial classifications – and they tell us little about what is happening within these categories. The Covid-19 pandemic has brought about a haemorrhaging of jobs in the airline industry, aircraft production and hospitality with many companies planning redundancies, but the longer-term effects on the structure of the labour market are as yet unknown.

The Blurring of Manufacturing and Services: Dispersed Surplus Value

One broad change in capitalism in general, as it has developed, is the expansion of the service sector in comparison to manufacturing. Advances in technology mean that goods can be produced with fewer workers and the inventiveness of capitalism offers an ever-increasing array of services. Therefore, over a period of time labour has been expelled from manufacturing and reabsorbed into jobs in the service sector. But it is worth noting that manufacturing has never constituted an absolute majority of the workforce – in fact manufacturing employment peaked at 40 per cent of the labour force in 1911 and even then was slightly smaller than the service sector (Hardy and Choonara, 2014). Since 1961 the gap between the proportion of people working in manufacturing and the 'service' sector in Britain has accelerated. Marx was clear that workers, whether making things or providing a service by working in a gym or restaurant, generate surplus value. In his own words, whether someone 'has laid out his capital in a teaching factory, instead of a sausage factory does not alter the relation' (1977: 477).

The distinction between manufacturing and the service sector has been blurred by the way many manufacturing firms have responded to competition by transforming themselves into service or part-service firms in what management speak calls 'product-service systems'. Over a third of large manufacturing firms offer services, including IBM, Rolls Royce Aerospace, Siemens and Xerox, which allow them to compete in their core markets affected by weak demand, intense competition and decreasing profit margins. The outsourcing of cleaning, computer programming, transport and a whole range of other functions previously carried out in-house, and now delivered by independent companies, inflates the scale of the shift from manufacturing. In addition, the economic value of manufactured goods increasingly depends on activities that are officially categorised as belonging to other sectors of the economy (Hauge and O'Sullivan, 2019). In other words, the production of surplus value is not just at the point of physical production; as competition takes on more sophisticated forms specialist functions, including research and development, analysis and testing, industrial design and intellectual property and distribution, take on more importance. As we shall see later this idea of dispersed surplus value is important for understanding why the distinction between 'knowledge' and manual workers is an artificial one.

The Slippery Slope of Relative Decline

Although it is the case that crude accounting underestimates the manufacturing sector in Britain, its historic decline in competitiveness means that this trend is more marked when compared to rival capitalist economies. Table 3.2 reveals a picture of Britain's relative decline in manufacturing and its ascendancy and consolidation as an exporter of commercial services (wholesale banking, insurance, currency trading).

In 1948 Britain accounted for 11 per cent of world exports; by 1980 it had slipped into sixth position, accounting for 6 per cent of world exports, and by 2019 the economy accounted for only 3 per cent of global exports and had fallen further down the league table to tenth position (Ward, 2020). However, Britain's position in exporting banking and financial services is a different story. In 2018 it was second in the global league table with a 6.5 per cent share of global exports in the financial sector reflecting its pivotal position in recycling the profits of global capitalism (World Trade Organization, 2019).

Table 3.2 Percentage share of global exports in merchandise and commercial services, 2018

Merchandise (goods)			*Commercial services*		
Ranking	*Country*	*Percentage share*	*Ranking*	*Country*	*Percentage share*
1	China	12.8	1	United States	14.0
2	US	8.5	2	United Kingdom	6.5
3	Germany	8.0	3	Germany	5.6
4	Japan	3.8	4	France	5.0
10	United Kingdom	2.5	5	China	4.6

Source: World Trade Organization (2019: 100 and 201).

Even between 1880 and 1914 Britain was already suffering major competition from industrial rivals in Germany and the US. In the first two decades after the Second World War the serious weaknesses of British capitalism were concealed by the prosperity and buoyancy of the recovering Western economy as it survived on the crumbs from the table of the economic miracles of Germany and Japan. But by the mid-1960s this spiral of relative decline was clearly evident. The growth rate in Britain between 1950 and 1973 was only 2.5 per cent compared with 4.9 per cent in Germany and 8 per cent in Japan (Griffiths and Wall, 2004). This was as a result, at least in part, of a failure to invest in machinery and factories on the same scale as other major European countries and Japan.

Three factors precipitated Britain's industrial decline in the ten years from the mid-1970s: entry to the European Common Market (the precursor to the EU), the world crisis of the mid-1970s and the monetarist policies of the Conservative prime minister, Margaret Thatcher. In the early 1970s Britain's final admission to the Common Market exposed its inefficient capitals to ferocious competition. But it was the crisis of the mid-1970s that opened up a new phase in the world economy and exposed Britain's eroding competitiveness, with the government bailing out the British Leyland car manufacturers, shipbuilding and British Steel to slow down their decline. The recession of the early 1980s resulted in a serious bloodletting as inefficient firms went bankrupt, and there was a loss of nearly 19 per cent in manufacturing jobs in four years from 1981. Margaret Thatcher's Conservative government was blamed for the dramatic deindustrialisation of the early 1980s as monetarist policies drove interest rates up to unprecedented highs and pushed the exchange rate of sterling to levels that made the products of firms uncompetitive. But rather than the Thatcher years being the cause of manufacturing decline and deindustrialisation, what they did was accelerate a process that had been underway for decades.

Real Production Still Matters

Although the decline of industrial production in Britain has been greater than in other advanced capitalist economies there has not been a collapse of manufacturing. Britain is still a significant exporter of goods, though there are significant variations in the profitability and dynamics of different sectors of the British manufacturing industry. By far the biggest contributor (by revenue) to total production is the food and drinks industry, dominated by large transnational firms. However, other industries – car production, defence, aerospace and pharmaceuticals – are hailed as the 'jewels in the crown' of British capitalism. In 2019, firms such as GlaxoSmithKline, AstraZeneca, BAE Systems (British Aerospace) and Rolls Royce were among Britain's largest manufacturing companies.

Britain has fared well in the interstate rivalry for foreign investment in vehicle production. In 2018 it was the fourth largest manufacturer in Europe and thirteenth largest in the world. In the same year the industry contributed 0.8 per cent of the whole economy's output and 8.5 per cent of manufacturing. The 166,000 employed in the industry manufactured 1.6 million cars: of these 80 per cent were exported (Rhodes, 2019). Inefficient capitals have been cleared out, usually in established older sites, in favour of greenfield production in new locations largely based on Japanese capital attracted to low-wage areas of Britain: the West Midlands and North East regions of Britain accounted for 43 per cent of jobs in the sector. At the time of writing the announced closure of the Honda factory in Swindon in 2019, with a loss of 3,500 jobs, signalled a much more uncertain future for the industry. However, despite dire prognostications from some Nissan's CEO announced that the Brexit deal from 1 January 2021 was good for Britain's car industry as exports between Britain and the EU are tariff free. The adaptation of British and foreign capital in the car industry to producing hybrid and electric cars is likely to be decisive for its future in European and global markets.

The pharmaceutical industry is hailed as one of the great success stories of British capitalism with two British firms, GlaxoSmithKline and AstraZeneca, occupying second and third place in the global league table of companies. The sector contributed around 1 per cent of the UK's output and 7.7 per cent of manufacturing. Employing 62,600 people directly and another 51,000 in related activities (BEIS Committee, 2018) it fits with the image of Britain as doing 'high-tech' production based on innovation. Similarly associated with high-level skills and good jobs is Britain's defence and aerospace industry, one of the largest in the world. However, it is heavily intertwined with and dependent on the national state as a major customer and to lobby on its behalf to secure orders with other governments.

Low productivity is the Achilles heel of British capitalism, historically trailing behind that of its capitalist rivals. Since 2008 it has stagnated and by 2019 it was only 2.9 per cent above what it had been twelve years earlier (Harari, 2020). In 2016 UK productivity was 16 per cent lower than the rest of the G7 economies[3] and lagging behind Germany, the US, France and Italy (ibid.). However, there is unevenness between the productivity of different sectors, and stagnation in some sectors and firms can sit alongside dynamism in others that are more successful at reorganising their capital under the whip of competition. For example, car production is the exception, outstripping the rest of manufacturing in terms of productivity. Therefore, pronouncements of the death of Britain's manufacturing sector ignore the existence of strong pockets of activity where their economic importance is disproportionately greater than the amount of employment generated. The statistics supporting the story of industrial decline underestimate the continuing importance of manufacturing by taking a narrow view of the chain of activities that result in the production of tangible goods and the way that surplus value is dispersed throughout this chain.

THE RISE OF THE ROBOTS

The Fourth Industrial Revolution

Spurring apocalyptic visions for work, some are claiming that current technological developments in contemporary capitalism are a uniquely disruptive force with no historical precedent. The term 'the fourth industrial revolution' has been coined to describe breakthroughs in artificial intelligence, robotics, the 'internet of things', autonomous vehicles, 3-D printing, nanotechnology, energy storage and quantum computing (Schwab, 2015). It is proposed that these developments are distinct from previous technological revolutions in their velocity, scope and systems, with Frey and Osborne (2017) claiming that in the US 47 per cent of jobs will be replaced by artificial intelligence and other new technology.

It is argued that the enhanced senses and dexterity of robots allow them to do the non-routine 'knowledge' work of skilled workers and that there is therefore a fundamental shift in the relationship between workers and machines with machines themselves turning into workers (Ford, 2016). Without minimising the achievements of scientists and engineers who have pioneered groundbreaking technologies that have liberatory potential for work and welfare, claims about their use and impact on work have been exaggerated. To begin with, other researchers have a much more sober assessment of the impact of recent technology, arguing that, although devastating for those workers, it will affect only 10 per cent of jobs as much automation has already taken place (Arntz et al., 2016). We are told that 'this time it is really different', but machines replacing workers is not new – recent technology is a continuation of processes begun in the industrial revolution. These changes were seismic in the social and economic upheaval they produced, with dramatic changes to the lives of workers as towns and cities grew exponentially. Neither is the debate about whether machines can replace humans a recent phenomenon; this was the subject of a paper published by Alan Turing in the British journal *Mind* as early as 1950.

Artificial intelligence (AI) is widespread in popular culture as robots are anthropomorphised in numerous films, even traceable back to Fritz Lang's *Metropolis* in 1927. But despite futuristic science fiction movies where robots learn how to think, feel, fall in love and subsequently take over humanity, the development and use of AI is more mundane and already deeply embedded in many aspects of everyday life. AI is simply machine learning. Every time you input data into your phone, or use apps or Facebook, these devices learn how to respond. Behind all the furore surrounding AI is a massive, insecure, low-paid workforce powering the wheels of giant companies like Google, Amazon and Twitter. The apparent alchemy of new technology masks the labour process and social relations underlying its production in what Marx referred to as the reification of commodities. Tedious and repetitive work underlies the practices of doing machine learning. There is an army of workers – collecting, cleaning and curating data; choosing or designing algorithms; altering code based on outputs; rating pages and filtering porn – often recruited from global sites such as Mechanical Turk, who do not appear on company payrolls (Pettersen, 2018; Wajcman, 2017).

Even though in Britain there is an increase in the use of robots in agriculture, manufacturing and even the domestic sphere, there is alarm among the custodians of British capitalism that the problem is not that there are too many robots, but that there are too few (BEIS, 2019). Industrial robotics have been a major contributor to the growth of labour productivity and value-added productivity across industrialised nations, accounting for as much as a tenth of aggregated growth between 1993 and 2007 (ibid.). But in 2015 Britain was lagging far behind comparable capitalist economies with only ten robots for every million hours worked, compared with 131 in the US, 133 in Germany and 167 in Japan, and by 2017 the UK represented just 0.6 per cent of industrial robotics shipments (ibid.)

The use of highly advanced computers by 'knowledge' workers and researchers does not reduce the social nature of work and the collaboration necessary for innovation. For example, in the games industry (discussed in Chapter 10), although the technology for their development is highly sophisticated the production of a game requires teamwork in conception and execution, which demands that developers work and communicate in the same physical space. Neither have the use of robots been a panacea in automobile production, where they have been widely used. Their inflexibility has led to some firms replacing robots with humans, or readjusting complete automation with a different work process.

Capitalism, Technology and the Dimension of Time

Beyond overarching macro changes to the structure of employment, new technology has been used in a specific and targeted way by capital to compress time in the process of the production and circulation of goods – what Marx referred to as the 'annihilation of space by time' (1973: 524).[4] For individual capitalists it means bringing goods to the market more quickly: eliminating waste in the form of time, effort, defective units and stocks in the manufacturing-distribution system to speed up circulation and the realisation of surplus value before the start of another round of accumulation. Food retailing, in particular, has been subject to constant change and transformation through a series of innovations. Just-in-time deliveries that reduce the time from field to shelf and systems that monitor spending habits are critical elements in the turf wars of large supermarkets.

The explosion of online consumption for nearly every type of good has reshaped the nature and geography of work. Between 2009 and 2019 the number of workers employed by Amazon in Britain has increased from 1,872 to 22,073, concentrated in 17 fulfilment centres (Statista, 2020). Online purchasing coupled with the domination of food retailing by a handful of big players and intense competition based on just-in-time delivery has transformed the physical geography of capitalism with the proliferation of gargantuan warehouses. In Britain between 2007 and 2018 about 235 million square feet was leased or purchased for warehousing – equivalent to 3,000 Wembley Stadiums – which represents a near doubling of demand for warehousing space in this period (Wearn, 2018).

Just-in-time is a weapon in competition between firms in their production and distribution, but it is simultaneously a weakness. A strike by those delivering components would bring production lines to a halt in a matter of hours. Industrial action taken by drivers delivering food to supermarkets would mean empty shelves in a short space of time. There is a concentration of workers in these new giant workplaces who have the same potential to talk, argue, share grievances and organise on questions of pay and working conditions as did the factories of the nineteenth and twentieth centuries. In one such warehouse, Sports Direct (featured in Chapter 10), the unionisation of 4,000 workers has been a challenge.

Filling the 'Pores of the Working Day'

Marx distinguishes between the extensification of labour by extending the working day (creating absolute surplus value) and the intensification of labour (relative surplus value) where workers are coerced into working harder in a given period of time. Longer working days have been constrained by legislation and hard-fought battles by workers, and therefore employers have resorted to intensifying work, often by using technology, in what Marx describes as filling up the 'pores of the working day' (1977: 386). However, while technology-related work intensification and measurement has a long history, digitalised communication technology takes this into new realms as work output and speed can be monitored remotely and extended to all types of work. Supermarket workers are measured by the number of swipes of barcodes, the number of calls taken by workers in call centres are monitored and every movement of delivery drivers can be tracked – even toilet breaks are measured. New levels of digital monitoring, which underpin the neoliberal workplace agenda with its emphasis on targets and competition, have been extended to the public sector – teachers and social workers included. Under pressure from slashed budgets for adult care, electronic monitoring has been used to strip 'unproductive labour' out of care work by surveilling the work of women carers to the last minute, while those working in community mental health are audited and monitored by electronic data dashboards (Moore and Hayes, 2017; Moth, 2020).

This technology is not the cause of neoliberalism but it has been harnessed in the workplace to deepen its metric-driven approach. The use of emails has ratcheted up stress by blurring the lines between work and personal life, while the expectation of a speedy reply keeps workers glued to their phones and computers and enters every crevice of their working lives. For many, working at home during the Covid-19 pandemic has increased the intensity of work as workers report an increase in micro-management. We might associate the use of electronic technology with 'professional' workers, but young workers in the hospitality sector interviewed for the Sheffield Needs a Pay Rise project and campaign reported that social media apps, such as Facebook, were used to contact them. The same apps functioned as a form of disciplinary surveillance as numbers of workers were reprimanded for venting against their employers on social media – even though these posts were heavily veiled.[5]

Knowledge and Creative Workers: Reasserting the 'Material'

There have been declamations, particularly dominant in the Italian autonomist tradition, that knowledge work and knowledge workers are replacing manual labour. It distinguishes between the traditional working class who work with their hands and bodies (material labour) and a new breed of worker who work with their minds (immaterial labour) (Lazzarato (1996: 133). In *Empire*, Hardt and Negri (2000) sketch a new era not only based on knowledge, information and communication but also on an emotional response – labelled affective labour. However, rigidly demarcating labour as a shift from the 'exploitation of *bodies* of workers during the Fordist mode of production to exploiting the *minds* of workers' (Woodcock, 2017: 55) in contemporary capitalism introduces an artificial distinction. The 'post-Fordist worker', as I argue in Chapter 4, is a glib generalisation from the experience of a small section of the labour force in the post-war period. Some services are deemed to use immaterial (as opposed to material) labour and draw on cognitive labour – what you do inside your head rather than with your hands. But many services either relate directly to the physical (material) world or are commodities themselves. For example, a software engineer might work on code to enable selling online, but at the end of the process there is a commodity. To see knowledge workers as a hermetically sealed group fails to recognise their role as directly or indirectly part of the process of producing material goods: digital games, medicines and engineering are part of the dispersed process of creating surplus value and the exploitation intrinsic to it. The idea that affective work is a new totem of contemporary capitalism is gender blind as emotional labour has always been associated with jobs in which women dominate – for example health and social work, education and roles such as human resource management. The idea that knowledge is the preserve of a prescribed group of researchers and scientists denigrates the skill and knowledge essential to occupations regarded as more lowly such as care work, as we shall see in Chapter 7, that undervalues what women do and the content of their work.

The label of creative industries is an attractive narrative that fits with the weightless economy and work that is innovative, dynamic and more satisfying. But the claim of a report by the Department of Culture, Media and Sport (2016) that in 2014 1.8 million people were employed in the creative industries sector, accounting for 8.2 per cent of the British economy, was creative accounting in itself. The calculations were inflated by including IT, software and computer services that constituted 43.5 cent of creative industries (gross value added) (ibid.). In fact, with the exception of IT, sectors that appear on the list of 'creative workers' have been around for decades (advertising and marketing, film, TV and video) while many others have existed for centuries (publishing, crafts, architecture, music). Being assigned creative capacity must ring hollow with museum and art gallery attendants taking tickets or standing for hours in galleries on minimum wage. In 2020, during the Covid-19 pandemic, this low-paid group of workers – in cleaning, retail and security in the commercial wing of Tate galleries – were regarded as expendable and made to take the brunt of the temporary closure as 313 jobs were threatened.

British universities are in the front line of the knowledge economy and the marketisation of these institutions in the last two decades (see Chapters 2 and 11) has reduced the production of knowledge to a tick box culture where narrow metrics discourage free and creative thinking. With 83 per cent of jobs in the pharmaceutical industry classified as research-based, successive governments and some academics hail the industry as a glowing example of 'high-tech', 'knowledge-based', 'research-oriented' jobs that are part of the 'new economy'. It rings all the right bells. However, science and research workers are subject to all the same pressures as other workers. There have been rounds of redundancies as firms close or merge. Work is intensified and the plug pulled on projects as these giant firms look for the next money-spinning drug (Randle, 1996). Tightening management control, cost cutting and reduced career opportunities were the lived experience of work for these knowledge workers.

Exploitation is an abstract concept, but the subjugation of both the bodies and minds of workers to capital has deep and far-reaching corporeal consequences. The interconnectedness of the physical demands of work and experience of alienation renders the distinction between manual and mental work unhelpful. Lack of autonomy, bullying and dismissive and disrespectful treatment by bosses spans both manual and non-manual work alike. Technology has always been a frontier of struggle. Dockworkers organised and took action against the containerisation that threatened jobs in the early 1970s. So-called modernisation was at the centre of the Wapping dispute in 1986 when Rupert Murdoch set up a new newspaper plant in East London, replacing hot metal with computer-based digital compositing and typesetting. Technology and scientific progress are neither automatically good nor bad – and neither are they neutral. The impact of technology on human progress in general and on working lives specifically depends on who owns and controls technology and how it is organised. In contrast to apocalyptic visions of total management control of the workplace, workers have used web-based communication to organise and fight back against the power of capital, as we shall see in the disputes covered in later chapters.

LURCHING FROM CRISIS TO CRISIS

Narratives about British capitalism, both by the ruling class and some on the left, need to be put under scrutiny. Categories that label workers as 'creative', 'knowledge based' or 'informational' potentially serve to create divisions in the working class and obscure the labour process that lies underneath. The structure of employment, which is constantly changing, is influenced by how competitiveness plays out between rival capitalist economies, state policies and the technological and managerial innovations of capitalism. But these are not linear and predetermined paths; they are interrupted, mediated and halted by workers' resistance. The system is punctuated by endemic shocks and crises that bring profound and often violent changes to the lives of working-class people. The fallout of the cataclysmic Covid-19 pandemic is not yet clear in terms of the deep structural change that will be brought about, a topic discussed in Chapter 13. However, it is now possible to see the patterns that have emerged more than ten years after the economic crisis of 2008. Although full-time work remains the norm for most workers, two-thirds of the growth in jobs has been in atypical roles such as self-employment, ZHCs or agency work (Clarke and Cominetti, 2019). Not only is this insecure work at the root of poverty and economic hardship, but it is argued that this group of workers are less able to defend themselves in a harsh labour market. It is to this debate that we turn in Chapter 4.

4

New Icons of Work? The 'Gig' Economy and Precarious Labour

A NEW WORKING-CLASS DIVIDE?

The liberalisation of postal services in 2011 opened up a 'wild west' in parcel delivery with a race to the bottom in cost cutting and speed of delivery. Alexandra, a Romanian migrant worker, described to me working for DHL on a self-employment basis (no sick or holiday pay and no pension) – the only contract on offer. Working a minimum twelve-hour day, he showed me the 100 drops on his phone (a drop can be up to four parcels). Even working these punitive hours at £1.85 a drop, Alexandra said that he is lucky to make the minimum wage after van hire, insurance and fuel have been paid.

One of the major claims about the changing nature of work that emerged at the beginning of the twenty-first century is that it has become more precarious. Along with the advent of the 'gig' economy, the growth of ZHCs and temporary work, and the use of employment agencies, are heralded as icons of the neoliberal era. The human cost of precarious work was illustrated by the tragic and unnecessary death of Don Lane in 2018, a DPD courier with diabetes, who died after missing hospital appointments with a specialist because he feared the draconian £150 fine from his employer (Booth, 2018). The exposure of notoriously poor working conditions in the Sports Direct warehouse and a spate of newspaper articles highlighting the scourge of ZHCs prompted the Conservative government to commission the *Taylor Review of Modern Work Practices* (Taylor et al., 2017). The report was greeted by employers with squeals of fright at the possibility that higher costs might be incurred in improving the conditions of employment, and was damned by trade unions as worthy but toothless.

The most extreme version of the argument of a growing underclass, peripheral to mainstream work and disconnected from organised labour, is encapsulated in Guy Standing's book *The Precariat*. He defines non-precarious work as including 'workers in long-term, stable fixed hour jobs, with established routes of advancement, subject to unionization and collective agreements with job titles their mothers and fathers understood, facing local employers whose names and features they were familiar with' (2011: 6). Standing goes on to describe what he calls the 'salariat' as those who are 'still in stable full-time employment, some hoping to move into the elite, the majority just enjoying the trappings of their kind, with their pensions, their paid holidays and enterprise benefits, often subsidised by the state. The salariat is concentrated in large corporations, government agencies and public administration, including the civil service' (ibid.: 12). In parallel to this decreasing group of privileged workers, Standing argues that there is a growing army of workers on insecure contracts, without benefits and without prospects for promotion. This is extended to suggest that rather than a job for life everyone will have lots of different jobs over a lifetime and often several jobs at once.

No one can be in any doubt about the material, social and psychological hardship that insecure employment practices inflict. Workers are condemned to living from one week (or even day) to the next, not knowing if they can make their rent and robbed of a social life as the demand that they are constantly available for work means they cannot commit to plans with family and friends. But this simple binary division of a group of workers with proper jobs and a growing army of precarious workers is not only overstated but also politically problematic. Coupled with amnesia about important historical struggles and a lack of attention to recent successful disputes, it encourages a defeatist mindset whereby the organisation of precarious workers by trade unions or their self-organisation is seen as a formidable or even impossible task. The collapse of the varied experiences of workers into the single category of 'precarious' by conflating different contracts masks the unique and specific exploitation faced by groups of workers in particular labour markets. For example, although a freelance fashion designer and machinist in a clothing factory are employed in the same sector and both have insecure work, that is all they have in common. While a designer may be able to command a reasonable fee and has some autonomy over their work, machinists are employed on low or poverty wages with no control over their labour process.

Proclamations that the number of those in precarious work is gathering momentum has lacked a thoroughgoing examination. While there are many accounts and testimonies that document the lived experience and hardships of insecure employment, interrogations of the data are few and far between. Definitions of precarious work have been fuzzy and the yardstick used to measure these apparently new employment practices is a relatively brief window of time – namely the three decades following the Second World War.[1] The stereotypical formula of lifetime employment with a pension, perks and job security only ever applied to some groups of workers – even in the 'golden years' of capitalism: women and migrant workers were largely excluded from relatively 'privileged' work.

This chapter begins by placing precarious work in historical context. I argue that British capitalism, from its inception and even through the post-war boom, relied on insecure work and a reserve army of labour, in some sectors at least. Evidence for the rise of more insecure work and the nature of the gig economy is scrutinised by looking at what the numbers tell us in general and by deconstructing the different categories of insecure work. The chapter tries to capture both the broad sweep of the neoliberal period and more specifically what has changed since the economic crisis of 2008.

OLD WINE IN NEW BOTTLES? A HISTORICAL PERSPECTIVE

Bourdieu (1963) is credited with coining the term précarité, using it in his research in Algeria to differentiate between workers with permanent jobs and those with casual ones. However, references to precarious work and ZHCs have appeared in common parlance in Britain only in the new millennium. This repackages a feature endemic to capitalism, which is that in parts of the labour market insecure work is the norm rather than the exception while many other sectors reveal a much more fluid relationship between secure and insecure workers. In the twenty-first century some workers wait for the next gig to pop up on their smartphone, but there is nothing new about casualisation and on-call arrangements, as we shall see in the next two sections.

Pre-1945: Casual Work Is Ubiquitous

In the nineteenth century social investigations by Charles Booth (East London), Seebohm Rowntree (York) and Eleanor Rathbone (Liverpool), among others, revealed that casual employment was ubiquitous and the norm. General labourers in seasonal trades (for example construction, dock and warehouse work, gas production and clothing) might be employed for just an hour or half a day (Whiteside, 2017). In these and many other industries a regular working week was virtually unknown, resulting in a 'festering mass of human wretchedness in all our great towns' according to philanthropist Sam Smith in 1883 (ibid.).

Writing about iron making in South Wales (Britain) in the 1830s, Williams (1978) illustrates how casualised and skilled work were often intertwined. He notes that there were 40 separate, hierarchically structured trades, each with different levels of pay and security: 'A striking feature of that working population was its sheer complexity and the strongly corporate spirit, which this often engendered among groups of workers who were self-recruited or organised by dozens of sub-contractors' (ibid: 43).

Nevertheless, despite this finely grained strata of occupations, the presence of subcontractors and the insecurity of the employment, workers from a 'bewildering web of trades' managed to come together in a massive insurrection that was brutally suppressed, but which led to the emergence of the organised working class in South Wales in 1831 (ibid.).

Nearly six decades later 'new unionism' was born in Britain from the Great Dock Strike of 1889, where for the first time unskilled workers organised themselves into trade unions. In the docks in London in the 1880s there were 150,000 workers dependent on work in ports. Only 10 per cent had permanent jobs – the rest would wait outside the docks on a daily basis to try to get employment. As men struggled to get a ticket to work, 'Coats, flesh and even ears were torn off ... mad human rats who saw food in a ticket' (Charlton, 1999: 32, citing Torr, 1956: 281). This was a fragmented workforce with intricate divisions of labour and where subcontracting arrangements prevailed. Despite these obstacles, workers organised a mass strike, which marked the beginning of the new unionism for unskilled workers in Britain.

Casual work and subcontracting were rife in parts of the coal mining industry, particularly in Derbyshire and Nottinghamshire where the 'butty system' operated. Mine owners divided the coalface into sections and parcelled it out to butties (subcontractors) who were paid by the ton to bring the coal to the surface (Johnson, 2015). The butty would employ men at a rate per day, 'pocketing the difference between his costs of production and his contract with the owner' (Rowe, 1923, quoted in Johnson, 2015: 5). The butty's income depended on 'the amount of drive which he could put into the men, the system involved much bullying, both moral and physical' (ibid: 5). The butty could select the day-wage men he wanted and therefore it was a system of nepotism and control in the local community. The butty held the sword of Damocles over families, even demanding sexual favours from the wives or daughters of men desperate for work (Johnson, 2015: 15). A miner wrote to the *Nottingham Journal* in 1922:

> The old butty system was responsible for a condition of things which should never be tolerated by any community today – bullying of men, which often led to accidents, bribery and corrupt practices of all forms to get hold of the best stalls – even a wife's honour has been sacrificed in thousands of cases in order to gain favour in this wretched system. (Cited in Johnson, 2015: 15)

Although the butty system nominally ended in 1919, it persisted in the Nottingham and Derbyshire coalfields, in different forms, for more than a quarter of a century. Workers agitated against the system and a big unofficial meeting was called at Mansfield Town Hall in 1919 with subsequent ballots in the pits showing massive opposition. As communists working in Pleasley mine pointed out, many of the union officials of the Nottingham and Derbyshire Miners' Associations were butties and therefore could not represent the interest of miners. The butty system in coal mining was only finally brought to an end with the nationalisation of the industry in 1945.

As we shall see in Chapter 7, the employment of women workers in the chain-making industry in the Black Country, north-west of Birmingham, at the beginning of the twentieth century could not have been more precarious. Earning not even a fifth of men's wages they forged chains in workshops adjacent to their home and worked for middlemen (subcontractors). Nevertheless, in 1910 this splintered labour force of women workers came together to demand the implementation of the wage recommended by the Trade Board and held out on strike for 20 weeks until all employers met their demands for higher wages (Barnsley, 2010).

Post-1945: Putting Standard Employment in Its Place

The ubiquitous casualisation of work in the period before the Second World War is hardly controversial. However, it is taken as given that the post-war settlement between capital and labour, brokered by the Labour government and continued by various Conservative governments, led to the so-called 'standard employment contract'. Undoubtedly, in comparison with the situation before the Second World War, there was a marked increase in stable work that carried benefits such as pensions and protections against unemployment. This was at least in part because a casualised and benighted workforce was not suitable for an aspiration to build a British economy that could compete with the economic miracles that were taking place in Japan and Germany. But secure work was not pervasive and there were plenty of sectors and jobs that did not reap these benefits.

Technical and craft work 'below the line' in the British film industry, dominated by large and profitable companies, was characterised by unpaid labour, uncertain employment and long working days (Atkinson and Randle, 2014). This industry did not have jobs for life, or a labour process that ever followed Fordist mass production lines. The term 'freelance' was a euphemism that dignified what now would be labelled casual or precarious labour. In their historical study of the post-war iron and steel, shipbuilding and docks industries Mankelow and Wilkinson (1998) show how employers, even in staple industries considered to be backbone of the economy, had a high degree of labour force flexibility. In iron and steel this was through promotion by seniority, which concentrated insecurity on the short-service workers and tied the economic interests of the long-service workers to their immediate employers. Shipbuilding and the docks were highly casualised, although job security was more equally shared.

The demand for labour on the docks in the post-war period continued to be irregular because of the uneven flow of shipping into the ports and cut-throat competition between employers. The response was a casualisation of the dock labour force that was organised around the call-on system whereby workers presented themselves for work each day and were selected if needed. Only a small proportion of dockers had permanent employment with one employer, and generally dockworkers worked for a large number of employers. In 1921 a registration system was introduced in London and although the average number of men required for work amounted to 34,000 per working day, the register included 61,000 men. Registration was a step towards controlling the chaos of the call-on system, but it did little to end casualisation, which remained the normal form of employment up to decasualisation in 1967.

Working practices in the building industry completely belie the notion of some universal golden age of secure work. It was notoriously difficult to establish trade unions on building sites as they were, by their nature, temporary, geographically dispersed and operated by different employers. To complicate matters further, on any given building site there were a large number of different trades, many in separate, small unions. When a building site finished, the process of negotiating new pay and conditions started all over again on the next one. In the mid-1960s employers were given tax concessions to allow them to promote 'the lump', a system where workers were hired on a self-employed basis and paid a lump sum for the work done each day or week. The incentive for 'working on the lump' was generous tax allowances for travel, lodgings and tools that were signed off by the employer so that workers paid lower income tax than a directly employed worker. This undermined the ability of trade unions to organise workers as many employers refused to take on direct labour, fearing that they might be trade unionists who could bargain on wages and health and safety.

The 1972 building strike, triggered by trade unions in the building industry demanding higher wages, exploded the myth that these workers were unorganisable (see Darlington and Lyddon, 2001, for a full account). A claim was put to the National Federation of Building Trades Employers for £30 per week and a basic 35-hour week for all trades. The employers rejected these demands and a national strike of selective sites was called in June 1972. The Union of Construction, Allied Trades and Technicians called out 35 major 'prestige' sites operated by major companies. According to building worker Gerry Kelly, rank-and-file pressure and action turned the selective strike action called by the union leadership into a nationwide revolt. A Strike Action Committee sprang up and spread the strike unofficially with flying pickets. By August 1972 there was an all-out national strike and the union leadership was left with no choice but to make it official (see Figure 4.1). In September 1972 the union side in the negotiations agreed a deal with employers to immediately increase basic rates of pay by £6 per week for craftsmen and £5 per week for labourers – the largest single pay increase ever negotiated in the building industry.

Figure 4.1 Demonstrating in defence of Shrewsbury building workers

Note: Five months after the building strike of 1972 ended, 24 North Wales pickets were charged with over 200 offences, including unlawful assembly, intimidation and affray. Six of the pickets were also charged with conspiracy to intimidate.

Source: John Sturrock.

It was a magnificent victory for trade union organisation, against the odds. The building trades employers, and their supporters in the Conservative government, were shaken because one of the least well-organised groups of workers had taken on their employers and won. But there was a backlash as employers and the state went on the offensive to try to break the high level of workers' confidence and undermine future action. In politically motivated trials in 1972, 1973 and 1974 eight members of the Shrewsbury 24 building workers were convicted on trumped-up charges and jailed for picketing (*Socialist Worker*, 2020).[2] The blacklisting of building workers was rife as the largest construction firms exchanged information on union activists, many of whom never got jobs in construction again (Smith and Chamberlain, 2015). The fight for compensation for ruined lives was only eventually successful in 2016 when millions of pounds was paid out by construction companies.

Women and Migrant Workers

Inasmuch as stable jobs with pensions, perks and employment security existed these were often, though not exclusively, the preserve of white, male workers employed by large organisations in manufacturing, engineering and chemical plants. Women and migrant workers did not have the same access to these relatively 'privileged' jobs. The lack of statutory entitlement to maternity leave until 1973 left women exposed to job insecurity. In fact, although there were growing numbers of female civil servants in the post-war period, it was not until 1973 that the marriage bar was finally abolished by the Foreign Office – the last bastion of misogyny. In the post-war period unregulated and hidden homeworking in the garment industry, and even assembling electronics, was still an important source of income and work for women, although the numbers involved are unknown due to the erratic collection of statistics and clandestine nature of the work.

Migrant workers in Britain were recruited to carry out low-paid and onerous work that native workers eschewed. However, this varied geographically. In multiracial and diverse places like the city of Birmingham and the industrial Black Country to its north-west, for example, migrant workers were much more integral to the large manufacturing and engineering workplaces, but, as I discuss in Chapter 8, they were still subject to systematic racism and discrimination. As Linda McDowell (2013 and 2016) shows in her rich ethnographic studies, migrant women workers were often on the bottom rung of the labour market, concentrated in insecure and invisible sectors of the economy such as cleaning or in small sweatshops. Even where they worked in skilled jobs such as nursing they would be concentrated in lower grades.

In the hype around the insecurity of employment associated with the gig economy, precarious work that prevails in a resurgence of 'sweated labour' in Britain has been neglected. 'Sweatshop' conditions of work in the garment industry are usually associated with developing countries, most notoriously Bangladesh, highlighted by the collapse of the Rana Plaza factory in Dhaka in 2013 where 1,134 women lost their lives and 2,500 were injured. However, many of these appalling conditions have been replicated in Britain. Leicester has the second largest concentration of garment manufacture in the country with 700 factories employing 10,000 workers. Although many of the factories are compliant with employment law, at the bottom of the supply chain are myriad small factories behind the veneer of terraced houses employing 10, 15 or 20 workers – often transient migrant labour on short-term visas. Work practices, in both registered factories and smaller unregulated production units, in Leicester are characterised by inadequate health and safety standards, verbal abuse, bullying, threats, humiliation and a lack of toilet breaks. Shop floor management often arbitrarily abuse and humiliate workers in front of co-workers for mistakes or missed performance targets (Hammer, 2015; O'Connor, 2018).

The history of capitalism is a story of the rampant and iniquitous casualisation of work in the period before the Second World War. But, an eagerness to mark out and emphasise the neoliberal period as one characterised by the resurgence and inexorable rise of casual work overstates the pervasiveness of the standard contract in capitalism's 'golden years'. The next section moves from narratives to numbers to disentangle the data on casual work.

DESTANDARDISING WORK IN THE NEOLIBERAL PERIOD?

The 'Gig' Economy

The most recent hyperbole about new working practices in the British economy centres on the 'gig' economy. In the broadest sense this is characterised by the prevalence of short-term contracts or freelance work akin to a 'gig' – a word originally coined in the 1920s by jazz musicians for a one-off engagement. Much of the current interest lies in the use of mobile application platforms (or app-based platforms for short) enabling work to be doled out in small chunks on the basis of piece rates. This makes deliveries, driving passengers or even cleaning homes akin to individual gigs associated with the entertainment industry. In 2017 these jobs fell into four main categories: courier services (City Sprint, Amazonflex); jobs obtained through website applications (Task Rabbit, PeoplePerHour); transport using own vehicle (Uber, Hailo/Mytaxi); and food delivery (Deliveroo, Ubereats). Although heralded as being technology driven, not all gig economy roles are based around a platform as workers may also work for more traditional companies that have simply imposed new contracts.

Using epitaphs of the 'sharing' or 'collaborative' economy and citing the supposed advantages of flexibility and autonomy, some have tried to dress these new ways of working as being benign and liberating for workers. The reality is that the vast majority of those in the gig economy work for profitable transnational corporations who, masquerading as 'technological companies', renounce their responsibilities as employers to unapologetically drive down their costs. The main mechanism for doing this is to misclassify their workers as self-employed independent subcontractors, which then allows these firms to transfer all of the risk onto the worker by denying them the most basic employment rights.

In British employment law there are three tiers of employment status. In the first tier, workers who are classified as 'employees' have job security, family-friendly rights, entitlements to sick and holiday pay, pensions and the right to a minimum wage. With no entitlements to any of these, self-employed workers are in the third tier. In the second tier lies the grey area of 'worker' legal status which carries some entitlements but no employment protection (sometimes known as dependent self-employed workers) (Böheim and Kepler, 2009; CIPD, 2020).[3]

Workers deemed to be self-employed are not only deprived of the rights and protections associated with employment, but at the same time miss out on the benefits of genuine self-employment, including the ability to control their own work. In the gig economy, despite the rhetoric of autonomy and flexibility, many of the traditional features of employment remain intact, with platform companies determining the conditions of the employment contract, including the ability to vet workers, the determination of rates of pay and how and when work is performed, requiring workers to wear a uniform with logos and even the ability to take disciplinary action. That is why in some cases it is not surprising that employment tribunals have ruled that an employment relationship exists and that individuals should be given the rights and entitlements associated with this.

Although the gig economy enabled by new technology is a fast-developing frontier of the labour market and of embryonic struggles, definitions are fuzzy and estimates of its growth and significance vary enormously. A *World Development Report* estimates that less than 0.5 per cent of the active labour force participates in the gig economy, with less than 0.3 per cent in developing countries (World Bank, 2019: 9). A Trades Union Congress (TUC) report (2019) claims that the gig economy in Britain has doubled since 2016, with one in ten working-age adults reporting working via gig platforms at least once a week, and argues that workers are increasingly likely to patch together a living from multiple sources. However, these claims need to be treated with caution. In 2016 around 1.1 million people worked multiple jobs, making up 3.6 per cent of the workforce, but this proportion is at an all-time low, having peaked at around 5 per cent in the mid-1990s (Corlett and Finch, 2016).

Despite the hype about technology, using apps does not always mean that all the activities they generate are new types of work. Work conducted through platforms is highly varied – it can, for example, include editing, design, software development and even legal and accounting tasks. These activities predated platforms by a long way and it may be that workers are using apps to do what they did before and the day-to-day content of their work has not actually changed. To include Task Rabbit and PeoplePerHour as examples of new ways of working enabled by apps is misleading. These are used by workers who are already self-employed – often in the building trade – and although auctioning their services online might intensify competition between them, tendering for work has always been part of the industry. In the contemporary economy the growth of precarious work cannot simply be attributed to technology. In fact, precarious work has grown more swiftly in well-established sectors that have been around for decades. Work in the 'gig' economy is nearly always precarious – but precarious work extends far beyond the use of platforms – witness the widespread use of these contracts in the care sector. The danger of fetishing the gig economy is to miss the fact that new technology is a fig leaf to conceal old methods of worker exploitation.

Deciphering the Numbers

Apocalyptic visions and narratives of the 'explosion' of precarious work are often painted with only cursory references to the numbers, or in other cases supported by playing fast and loose with the statistics. A report by the GMB trade union (2017) comes up with the 'staggering figure' that one in three workers are in insecure work – equivalent to 10.2 million workers. However, this figure conflates different categories of work – ZHCs, agency work and temporary and part-time work. This throws into the mix casual and temporary work that have always existed along with the relatively recent rise of temporary employment agencies and the relabeling of established practices such as ZHCs. Two contributions have put claims of insecurity in the neoliberal period under the microscope. Deploying data from the post-war period, Kevin Doogan's book, *New Capitalism*, dismantles 'smokestack nostalgia about proper jobs' (2009: 206) in advanced capitalist economies, while in the case of Britain Joseph Choonara's *Insecurity, Precarious Work and Labour Markets* (2019) conducts a forensic statistical examination to challenge the idea that there is a linear progression from secure to insecure work. The rigour of these analyses stands in sharp contrast to the flimsy evidence presented by Guy Standing.

One claim is that Britain's employment boom has been dominated by a growth in jobs in the gig economy and other types of insecure and atypical work. There is some truth in this claim. From 2008 to 2018, two-thirds of net employment growth has been in atypical employment (part-time work, temporary work, agency work or work on ZHCs) (a rise of 1.9 million out of 2.7 million jobs) (Clarke and Cominneti, 2019). Sometimes through choice, but often through the coercion of cuts to benefits, these jobs were taken up disproportionately by single parents, young people, people with disabilities and migrant workers. But as the labour market has tightened the growth in atypical employment slowed and then stopped, with all employment growth since 2016 driven by full-time employee work (ibid.). However, the terms atypical, non-standard and precarious work are overlapping and confusing. For example, part-time work is deemed to be atypical, but it may be done on a secure permanent contract. This category includes a large proportion of women workers. With childcare only affordable for the well paid and the burden of caring increased by public services slashed to the bone, this form of work is 'chosen' in order to juggle family and work commitments.

ZHCs – what used to be referred to as on-call work – are viewed as synonymous with precarious work. According to the ONS the percentage of people on ZHCs was a maximum of 0.8 per cent of the labour force between 2000 and 2012, after which it increased sharply and by June 2020 amounted to 3 per cent of the labour force – equivalent to 1.07 million workers (ONS, 2020b). Although the use of ZHCs appears to have increased, it is difficult to disentangle the actual growth of this type of work from the growth in awareness. The reliance on self-reporting these contracts in the collection of data may mean that individuals are simply rebranding working practices that they experienced before (ibid.; Choonara, 2019). Further, the use of these contracts is not widespread across all sectors, with the highest use of ZHCs in the hospitality sector (bars, hotels, restaurants), accounting for 22 per cent of employees in the sector. Traditionally this sort of work has always been highly casualised and therefore the figures reflect more continuity than change. More surprising is that ZHCs lie at the heart of the welfare state, accounting for 20 per cent of the workforce in health and social care. Even in education 9.5 per cent of workers are on these contracts (ONS, 2020b).

Between 2001 and 2017 self-employment increased from 3.3 million workers (12 per cent of the workforce) to 4.8 million (15 per cent of the workforce) (ONS, 2018). There was a sharp increase in the over 65 age group in this category after the 2008 crisis and a strong growth in the 16 to 24 age group. In the case of older workers this is often due to delaying retirement and continuing what they did before on a self-employed basis. But far from this signalling an explosion of 'entrepreneurialism', as we have seen a proportion of this self-employment is bogus, false or coerced, where employers have avoided giving workers their employment rights by labelling them as 'independent contractors'. While there are no accurate data, estimates are that around 460,000 workers fall into this category (CAB, 2015) – that is, 10 per cent of the 'self-employed'. A telling statistic is that between 2006 and 2018 the number of people moving from employed to self-employed status increased fourfold, while the number of people moving from unemployment to self-employment increased ninefold. This supports the argument that some groups have been coerced into the workforce because of draconian cuts to a range of disability benefits.

Recruitment agencies, particularly those that 'supply' temporary workers, are lucrative businesses that range from global players to micro firms that border on the illegal. This plethora of transnational corporations and 'one-man bands' was underlined when the Equality and Human Rights Commission found that as many as 131 agencies supplied workers on a temporary basis to the meat-packing industry (EHRC, 2010). The use of temporary agency workers grew from low levels in the early to mid-1990s to about 800,000 by 2018. Many firms use temporary workers to deal with fluctuating output, and sometimes for specialist professionals such as nurses, teachers or IT workers. But in other cases the use of temporary workers has been a deliberate tactic to drive down costs and instil fear and insecurity into workers. A disproportionate number of workers at the receiving end of this 'business model' are migrant workers, where a lack of knowledge of legal entitlements leaves them open to abuse. In Chapter 9 this is explored in the case of the Sports Direct warehouse.

Choonara (2019) found that temporary work has remained stable since the 1980s, and although this is more concentrated in the 16- to 24-year-old group, there was no evidence that it is feeding into greater temporary work later in life. Tenure across the labour force has remained remarkably stable, with the typical employee in 2015 in a job that can be expected to last on average 16 years – roughly the same as in 1975 (ibid.). Overall Choonara calculates that if all of these non-standard forms of employment are taken together they have expanded by 4 per cent of the labour force since the 1980s. Taking a shorter time frame, Clarke and Cominetti (2019) conclude that full-time work for an employer remains the norm and the structure of the labour market in 2018 was similar to 2008 with only a small decrease from 64.4 per cent in full-time employment to 63.1 per cent in the later period.

There are other factors that militate against the apparently relentless drive to insecure contracts. Not only do employers need labour, they need a trained and stable workforce. A high turnover of employees means the loss of embedded skills and knowledge in which bosses have invested; hiring new workers is expensive. This partly explains why, during the recession that followed the 2008 financial crisis, redundancies and lay-offs were proportionately less than the fall in demand and production. In the words of mainstream economists, bosses hoarded labour in anticipation of an economic recovery.

Feeling More Insecure

Measured by different types of contracts the numbers do not support the idea that there is an inexorable march from stable jobs to insecure work. But workers are facing new forms of insecurity and precariousness – both in terms of falling living standards and their experience of work. Perversely, in the decade after the economic crisis of 2008, and the austerity that followed in its wake, employment rose to unprecedented levels. However, earnings have not recovered from their 2008 level and as workers have faced a decline in their living standards they are having to work harder and longer hours just to stand still. A key driver of the rise in employment during the period 2010 to 2020, both in terms of numbers of people and hours worked, is the large shock to income that followed the financial crisis, and the glacially slow recovery of wages since (Clarke and Cominetti, 2019). Therefore, it is not the case that people have suddenly got an appetite for spending more time at work rather than with their friends and families or on their allotments – the decline in people's real pay packets has pushed up the average hours people have to work to earn enough to stay afloat.

A Resolution Foundation report by Clarke and Cominetti in 2019 celebrates this higher employment as more fairly sharing opportunity in Britain on the grounds that between 2007 and 2017 people living in households in the bottom half of the income distribution accounted for 62 per cent of the employment increase. In their words, 'the jobs boom has brought some of the most disadvantaged groups into employment. Ethnic minorities and people with relatively low qualifications have been the main beneficiaries, as have people with disabilities' (2019: 5). However, this benign view of the labour market as a place of opportunity quickly unravels. Rather than the labour market offering opportunity and choice for disadvantaged groups, many people have been coerced into the workforce through draconian legislation. In 2012 the UK coalition government introduced the harshest regime of conditionality and benefit sanctions in the history of the UK benefits system. Single parents, care givers and people with health difficulties were disproportionately vulnerable to, and affected by, the withdrawal of their benefits.

A new form of precariousness has been the rise of in-work poverty affecting one in seven working families in 2018. There are now equal numbers of households in poverty with someone in work as those where no one is in employment (Clarke and Cominetti, 2019). Reductions in benefits from the introduction of the Welfare Reform and Work Act in 2016 has left families with little protection to cope with low earnings as rising costs of housing take a greater share of their incomes. The use of food banks in Britain, almost unheard of before 2004, increased from 26,000 people in 2008–9 to 1.6 million in 2018–19 (Clark, 2020). In 2018 one in seven people using food banks were in employment, or lived with someone who was; often that work was part time (Trussell Trust, 2020). The fall in income of low-paid workers during the first two months of the lockdown during the Covid-19 pandemic pushed people into financial hardship. By the second half of March 2020 there was a 67 per cent increase in the number of household referrals to food banks and by April the number of children needing support had doubled (ibid.), a measure of the increased poverty and desperation into which people were plunged.

An additional source of precariousness, particularly in the public sector, lies in the intensification of work. This began in the private sector in the 1980s and soon spread to the public sector in the form of the 'New Public Management' developed in the Thatcher period in the 1980s to 'inject competition' and 'market forces' into the provision of government and local authority services. Subsequently these methods have been used increasingly, first in the civil service and more recently in schools, colleges and universities. The collaborative nature of much public sector work and the associated difficulties with measuring the output of workers has not stopped employers from imposing arbitrary 'performance targets' and 'metrics'. Therefore, people are rendered subjectively precarious by the increasing power that senior managers try to wield over them; the constant pressure to meet these demands coupled with a merry-go-round of reorganisations make people feel disposable. Figures from the Health and Safety Executive (2020) paint a stark picture of the workplace experience in 2018–19 with 602,000 cases of work-related stress, depression and anxiety reported. In 44 per cent of cases workload, tight deadlines, too much responsibility and a lack of managerial support were cited as the main reasons for stress.

A report by the *British Journal of Psychiatry* reveals a story of an explosion of anxiety. It notes that 'generalised anxiety disorders and symptoms began their upward trajectory around the time of the effects of the 2008 economic downturn during the policy of austerity' (Slee et al., 2020). In other words, stagnating incomes, deteriorating public services and the intensification of work have made people's working lives feel more precarious.

ALL WORK IS PRECARIOUS UNDER CAPITALISM

The ramifications of accepting the imagined division between secure and precarious workers is to see Marx's revolutionary class as a dwindling minority, a remnant of a previous era of job security and organised labour. But the image of a largely stable male, union card-holding and 'boilersuited' proletariat has always been the cartoon version, and precarity has been woven throughout working-class history. Nevertheless, victories have been scored against even the most unscrupulous firms. A lengthy legal battle by workers who were members of or supported by the GMB and IWGB to be classified as 'workers' not independent third-party contractors was won in a momentous victory in March 2021. This means that they are entitled to basic employment protection, including the minimum wage and holiday pay. While this is an important step, the fight continues for better conditions for Uber workers, including union recognition. This sets a precedent to end exploitative practices synonymous with gig economy firms and opens up a new arena of struggle.

There is not one, but a variety of regimes of insecure work and, as subsequent chapters show, precarious work is experienced in a variety of different ways. The women care workers in Birmingham and Glasgow, in Chapter 7, were primarily on part-time contracts – but these were permanent. The root of their insecurity lay in low pay that rendered the costs of day-to-day life a perpetual struggle. Similarly, the minimum wage paid to the (mainly) migrant London cleaners, who are the focus of Chapter 9, was compounded by their exclusion from entitlements to sick and holiday pay. Another dimension of precarious work is manifest in the 4,000 workers 'supplied' by two employment agencies to the Sports Direct warehouse, who are controlled and disciplined by the fear of having hours withdrawn.

Neither is precarious work restricted to low-paid workers: higher and further education have one of the highest rates of hourly paid and short-term contracts in the whole economy. In the 'salariat' of public sector employees, where full-time contracts prevail, the precariousness of work is characterised by extraordinary and rising levels of stress experienced by health workers, teachers and other local government workers under the strain of spending whittled to the bone by austerity and an intensification of demands on their time. The purpose of identifying the distinct exploitation that prevails in diverse sectors and jobs is not to compile a catalogue or pecking order of misery and oppression, rather it is to recognise and explore the different challenges, opportunities and strategies for organising and taking action.

The notion of precariousness is useful because it is intrinsic to all work under capitalism. In a system whose lifeblood is profits and competition, precarious work is inherent to all labour–capital relationships, with the continuous dislocation of work and workers through the mobility of capital and the impetus to intensify work and hold wages down. Beyond incremental changes in capitalism, insecurity is caused by its crisis-ridden nature dramatically illustrated in the first two decades of the twenty-first century by the eruption of two convulsions – the financial crisis of 2008 and the Covid-19 pandemic of 2020–1.

Those on precarious contracts are part of the working class and not a separate entity. Neoliberal capitalism, in the form of the state and capital, has waged a concerted attack to roll back the gains that a particular generation of workers was able to fight for. Those on temporary and part-time contracts have not got distinct interests from those in full-time and unionised jobs – they are often in the same sector or trade union. This chapter has looked at historic examples of where unorganised and fragmented workers have taken strike action from below and won under exigent conditions. Subsequent chapters look at how, since 2016, precarious workers have, despite barriers to collective action, come together to secure some stunning and life-changing victories.

5

Explosive Struggles and Bitter Defeats

THE HIGHS AND LOWS OF STRUGGLE

In 1972 Alan was a young engineer in Birmingham. He was a socialist and had been a shop steward in the factories he had worked in. Between jobs he had been on the picket line for ten days at the Saltley Gate coke depot trying to stop the movement of coal in solidarity with striking miners who were fighting against austerity pay deals. With a daily average of 200 pickets and 800 police, the lorries were getting through. On 3 February 1972 that changed dramatically when 30,000 Birmingham engineers walked out on strike in support of the miners and up to 15,000 of them marched to join the picket line at Saltley Gate. Alan describes how:

> Suddenly a sea of workers came over the brow of the hill. The feeling of power and solidarity was indescribable – there was a mood of jubilation. The chief constable of Birmingham contacted the prime minister and said that if they didn't close the gates, he couldn't guarantee the safety of Birmingham. The blockade forced the police, who had kept the depot open all week, to surrender. A huge cheer went up when they closed the gate.

Militant picketing involving tens of thousands of miners had shut down power stations, docks and coal depots. But the victory at Saltley, won through solidarity strikes, was the turning point for the miners. Within seven weeks Edward Heath's Conservative government was defeated. Twelve years on from this high point in British working-class history there was a bitter defeat when, in March 1985, miners' called off their strike against pit closures. In a bitter mood they marched back to work almost a year after the strike had begun. A delegate conference of the National Union of Mineworkers voted to abandon the strike without an agreement with the National Coal Board and called for an organised return to work. This heralded a grim period for the British working class.

Rosa Luxemburg understood not only how strikes struck fear into the hearts of capitalists as their profits are hit at the point of production, but also the trepidation of trade union leaders fearful that strikes might take on a life of their own and threaten their role as arbiters of capital and labour. In her pamphlet *The Mass Strike, the Political Party and the Trade Unions* (1906) she argues that a mass strike that begins with economic issues about pay, working conditions and victimisation can trigger wider political demands for progressive legislation and democratic rights and even spill over into challenging capitalism itself. For workers, going on strike is their most potent weapon, drawing on their unique position as the sole producers of wealth. It is a transformative, exhilarating experience, giving workers a sense of power and collectivity that throws the tyranny of work into sharp relief.

A 'Labour of Sisyphus'

Trade unions grew out of the collective response and fight of working people against poor wages and harsh working conditions. The struggle to organise met with punitive and aggressive responses from employers, supported by the machinery of the state. Many activists will relate to Rosa Luxemburg dubbing the work of trade unions as a 'labour of Sisyphus', referring to a figure in Greek mythology who was doomed to repeatedly roll a boulder uphill only for it to repeatedly crash to the ground. This metaphor captures the guerrilla warfare in workplaces where employers are constantly trying to undermine conditions of service and pay and move the frontier of control of the labour process in their favour. But it also refers to arguments and skirmishes within trade unions themselves as activists, particularly in some periods, try to overcome the lethargy and obstruction of union machinery in order to mount effective action against employers. Ambivalence in the function and actions of trade unions was captured by C. Wright Mills in his description of the trade union leaders as 'managers of discontent' who 'organise discontent and then sit on it' (1948: 8–9). Unions can be accommodated within capitalism to control and defuse anger, with negotiations serving to divert workers' grievances into stable channels and rob them of their potentially explosive content. But trade unions walk a tightrope. On the one hand, they are an institutional siphon for grievances, but on the other, if workers' action and discontent is dampened too far then the whole reason for their existence is fatally undermined. They operate in the system of capitalism, not seeking to overthrow it but to improve the position of workers within it; their aim is not to end exploitation but to renegotiate the terms on which workers are exploited.

The consciousness and fighting strength of the working class will vary between periods. Mass strikes have come in waves, in broad explosions of creative militancy, occurring erratically and unexpectedly. For example, in the UK there was the 1926 General Strike and the strike waves of 1934–9, 1968–74 and 1978–9, and in 2011 there was a strike by 2.5 million public sector workers (Darlington, 2014a). The changing balance of forces between capital and labour has a profound impact on the nature of trade unionism and its role and ability to mount resistance to exploitation, in terms of which side is more confident and stronger in pushing the frontier of control to their advantage within the workplace. This chapter begins with a broad outline of trends in strikes since 1945, arguing that any assumption that these figures signal an end to workplace conflict needs to be treated cautiously. The discussion goes on to look at how the key contradictions of workplace unionism, as a source of order or conflict and stability or resistance, has played out in different post-war epochs as successive governments have tried to contain and resist industrial action by workers.

THE SHAPE OF STRIKES IN BRITAIN: A HISTORICAL PERSPECTIVE

Four interrelated features dominate the last 50 years of strikes: there are far fewer of them; the main locus has shifted from manufacturing to public services and privatised industries; unofficial strikes no longer predominate; and courts and the legal system have been brought centre stage in the drama of industrial relations. Dave Lyddon (2015a) divides the pattern of strikes into three stages (see Table 5.1).

In the first phase, 1964 to 1979, the annual number of strikes fluctuated between 2,000 and 3,000. The most strike-prone industries in the 1960s were coalmining, shipbuilding, motor vehicles and the docks, which were characterised by highly fragmented rather than national bargaining, fluctuating earnings and significant bargaining opportunities at workplace level. On average 95 per cent of recorded strikes were unofficial through the 1960s and 1970s and in the mid-1960s about a third of them lasted a day or less. In the second phase the biggest recession in 50 years and the assault on trade unions and the working class by Margaret Thatcher's Conservative government in the 1980s brought about a transition from a high to a historically very low level of strike activity. In the third phase, from 1992 onwards, the annual number of strikes has been lower than at any time since 1893 (when the collection of statistics began).

Table 5.1 Officially recorded strike activity in Britain, 1964–2019 (annual averages)

Phase	*Period*	*Number of strikes*	*Workers involved*	*Working days lost*
Phase 1	1964–7	2,233	759,000	2,597,000
	1968–74	2,846	1,684,000	11,703,000
	1975–9	2,310	1,658,000	11,663,000
Phase 2	1980–4	1,351	1,298,000	10,486,000
	1985–90	838	702,000	3,600,000
Phase 3	1991–6	244	226,000	656,000
	1997–2001	192	145,000	357,000
	2002–14	131	514,000[a]	702,000
	2015–18	92	77,000	310,000

Note: [a] Figure boosted by national public sector strike on pensions.
Source: 1964–2014, Lyddon (2015a: 735); 2015–18, author's calculations from ONS statistics.

From September 1984 the law required ballots for union action in order for them to retain legal protection for official strikes. But unofficial stoppages still featured in the build-up to official action and even at the end of the 1980s three-quarters of strikes were still unofficial. In 1989 a government green paper cited high-profile unofficial strikes as evidence of this 'long-standing problem in industrial relations' (Lyddon, 2015a). This tipped the balance of power within unions to the longstanding preference (since 1945) of trade union leaderships to call national official 'token' strikes or discontinuous action and for selective rather than all-out strikes.

However, an overreliance on crude strike statistics can be a source of despondency for activists and triumphalism for the ruling class. These figures are the tip of the iceberg and fail to capture ongoing turbulence in the workplace. Shalev describes strike data as 'Some of the most abused and least understood of man's attempt to freeze and condense richly dynamic social events into static, artificial and misleadingly accurate arithmetic' (1978: 1). Case studies later in this book look at the anatomy of a series of disputes, each of them with different dynamics in terms of their triggers, their agency and the way that they were prosecuted. But strikes are a culmination of workers' anger and frustration; sometimes the tinder of grievances has built up over a long period of time, in other cases an unanticipated offensive by the employer provokes a quick response.

At a macro level the crude figures on strikes say nothing about the economic and political forces in different periods that shape the form of struggle and strikes. Chapter 4 looked at how competition between capitals and technology has reshaped the terrain in which exploitation is inscribed. The focus of this chapter is on how the state in different periods has tried to control and rein in the struggle of workers, through legislation and sometimes all-out confrontation. But the story is not simply one of labour versus capital, it is one of a shifting balance between trade union leaderships and their rank-and-file members and activists. Periods of trade union leaders opposing strikes have been interspersed by windows of time when they have been pushed into supporting strikes – even against Labour governments. These vacillations are the product of the Janus-faced nature of trade union leaders trying to balance maintaining credibility with both their members and with employers. This unstable and fluctuating situation is one where the balance of forces between employers, the state and capital does not remain in equilibrium but can change, sometimes dramatically, in different periods. Mainstream accounts of the struggles of workers are told in terms of conflict between trade unions and employers – where the former are treated as monolithic institutions. But an overview of different periods of time in the post-war era shows a much more complex cast of players in the drama of strikes and industrial action. The relative strength and confidence of the state, capital and trade unions, and within unions between ordinary members, the bureaucracy and their leaderships, plays out in different ways.

FROM POST-WAR 'GOLDEN YEARS' TO CRISIS: 1950 TO 1979

'I'm All Right Jack': 1950s to Mid-1960s

Workplace trade unionism in the 1950s has to be understood in the context of the shop steward, or workshop-based trade union representative system that had originated among skilled male workers in the shipbuilding and engineering industries at the end of the nineteenth century. They collected subscriptions and were the direct representatives of workers in their sections or 'shops', handling a wide range of issues with employers on their behalf. This steward organisation mushroomed during the First and Second World Wars. During the First World War a rank-and-file movement brought together shop floor representatives from different workplaces and unions to coordinate their struggles and at its height was able to organise significant strikes, the largest of which was in 1917 involving 200,000 engineers in 48 different towns (Darlington, 2014a). After 1945 the steward system spread to many industries and the direct representation of shop stewards/workplace representatives at the point of production was sustained. In the 1950s shop steward organisation and potentially independent trade union machinery had developed out of the Joint Production Committees formed during the Second World War. This workplace-based trade unionism and the frequent, but short, 'wildcat' strikes outside the official trade union machinery were condemned by the ruling class as the culprits of the poor performance of British capitalism. Blaming rising wages for inflation was a popular theme and workplace militancy and shop stewards were targets for widespread attacks by newspapers, film makers[1] and Conservative politicians throughout the 1950s.

The 1968 Donovan Report,[2] the Labour government's attempt at wrestling with the problem of unions that were deemed to be out of control, highlighted three main issues with British industrial relations. First, recorded strikes (outside of coal mining) trebled from the mid-1950s to the mid-1960s, almost all of which were 'unofficial' (lacking union approval) and in breach of an appropriate procedure agreement. Second, as trade unions could push up pay in individual workplaces the gap between negotiated industry-wide pay rates and actual earnings (including piece work bonuses and overtime) was growing. The third feature was that this informal system offered employers no means of negotiating and doing business with the trade union bureaucracy. Tight labour markets and full employment meant that workers could not be disciplined by unemployment and in the heartlands of manufacturing in North London and Birmingham a worker could walk out of one job in the morning and be working somewhere else by the afternoon.

Mounting Anger from New Trade Unionists: Mid-1960s to 1970

By the end of the 1960s the idea of a typical trade unionist as male (mainly white), boilersuited and working in a traditional industry no longer fitted with the changing economy and a new layer of activists. The anti-Vietnam War movement, emerging women's liberation and civil rights movements had politicised and energised not only students but large numbers of young workers. Unions were at the beginning of a great expansion that would add three million members between 1968 and 1979, including 1.5 million in public services – health, education and central and local government – and an extra 600,000 among white-collar workers in manufacturing that brought women into active trade unionism (Darlington and Lyddon, 2001).

Under the Labour government, in 1969 to 1970 there was a wave of disputes about pay among sections of workers – teachers and 'dustbinmen' – for whom trade unionism was new. Two strikes in 1970 – mainly women, clothing workers in Leeds and workers at the Pilkington glass factory in St Helens – were symptomatic of the underlying anger at years of pay restraint under Wilson's Labour government and frustration at their own trade union leaderships. The setting up of rank-and-file committees to organise the strikes presaged the explosion of rank-and-file workers' action that was soon to erupt. The politics of these two disputes are vividly captured in two plays by the BBC: *Leeds – United!* (1974) and *The Rank and File* (1971).

The Explosion of Workers' Struggles: Early 1970s

After growing discontent and the disputes of the late 1960s there was an explosion of workers' struggles in the early 1970s in unprecedented levels of militancy that brought down the Conservative government of Edward Heath in 1974. The overarching economic context of the convulsions of the early 1970s was the cumulative lack of competitiveness of the British economy in comparison with established competitors such as Germany and the rising stars of Japan and South Korea. In order to catch up and drive down costs there were few years throughout the 1960s and 1970s without some forms of restriction on wage increases (incomes policy). Another front in the conflict with workers as competitiveness waned was opened up by the threat to jobs either through the direct closure of workplaces or so-called modernisation. A rich and forensic account of this period, in terms of the scale and depth of militancy, the confidence and solidarity that it engendered and the pivotal role of rank-and-file workers, is captured in Ralph Darlington and Dave Lyddon's book *Glorious Summer* (2001).

On the election of a Conservative government in 1970 trade unions and workers faced a prime minister, Edward Heath, whose mission was to be even tougher on the working class. He took up the baton from the previous Labour government which had been forced to abandon its legislation[3] designed to shackle trade unions and destroy shop steward's organisation in the face of a mounting strike wave at the end of 1969. The nakedly anti-union Industrial Relations Act, introduced in 1971, was a clumsy piece of legislation that quickly brought the state into confrontation with workers. Initial success at quelling industrial action by workers was short-lived when watershed struggles of shipyard workers, miners and dockers erupted and forced the Heath government to do humiliating U-turns. In 1972 the imprisonment of five dockers under the Act for picketing provoked unofficial walkouts, initially from the dockers, that snowballed into mass strikes involving all workers across the country (see Figure 5.1). A previously unknown figure – the Official Solicitor – was wheeled out to provide a fig leaf for the complete retreat of the government. The five dockers were released and the Industrial Relations Act was dead in the water. The final showdown of the Heath government was another strike by miners. Amid power cuts and blackouts and a three-day week at the end of February, Heath called an election in February 1974 on the question of 'Who runs the country?' It turned out not to be him – a minority Labour government was elected. Although the triggers, details and prosecution of the strikes and workplace occupations were different, the element critical to their success was that fresh layers of activists wrested the control and tempo of disputes from the official union machinery. This massive upsurge in industrial action forced the leadership of trade unions to the left and the battles inside trade unions led to the emergence of a new generation of militants, radicalised by the politics of the late 1960s and more ready to challenge incumbent leaderships and find new ways of revitalising the confidence of the labour movement.

Figure 5.1 The release of five dockers in July 1972 imprisoned in Pentonville Prison for picketing under the Industrial Relations Act

Note: After unofficial solidarity walkouts snowballed into mass strikes across Britain they were released within a week.

Source: Socialist Worker archive.

Social Contract (Social Con Trick): 1974 to 1979

The Labour Party under Prime Minister Harold Wilson (and subsequently James Callaghan) was elected as a minority government in February 1974. In autumn of the same year a second election delivered a majority. Both the Industrial Relations Act and Heath's formal wage restraint were scrapped and the miners' demands were met virtually in full, but in its place trade union leaders agreed to the Social Contract. The architects of the deal, Jack Jones and Hugh Scanlon, considered to be left-wing trade union leaders, agreed to deliver voluntary wage restraint and police their rank-and-file members on behalf of the government.

However, the Labour government took office under the worst possible circumstances. Oil prices had quadrupled in 1973 and the global economy plunged into recession. In September 1976 the government had to go to the International Monetary Fund for a bailout to the tune of $3.6 million, the largest request for a loan in its history. It came with stringent conditions of slashing public spending to which the government quickly acquiesced. This was a watershed moment when Denis Healey (chancellor of the exchequer) sounded the death knell of Keynesianism that had purportedly formed the basis of the post-war consensus of both political parties.

Between 1975 and 1978 the Social Contract was successful in holding down wages in the face of rising inflation. However, in autumn 1978, when Prime Minister James Callaghan and Chancellor of the Exchequer Dennis Healey demanded another year of pay restraint, the dam burst and bitterness and frustration over falling real wages exploded in a series of strikes in the public and private sectors. Sensationalist images of the so-called 'Winter of Discontent' in 1978–9 peddled by the right-wing media were of uncollected municipal rubbish, union pickets at hospitals blocking the entry of medical supplies and the refusal of Merseyside gravediggers to bury the dead. But this is only a partial picture of the scale of revolt (for a full account see Lyddon, 2015b). There were strikes right across the public and private sectors, including unofficial strikes by Ford car workers and road haulage workers and 20,000 bakers on strike for five weeks in November and December of 1978.

The Labour government paid a high price for trying to foist the burden of the crisis of British capitalism onto workers by holding down wages and cutting public spending. The shift in policy from welfare and public spending to austerity and inflation and the trumpeting of market forces and finance led to Labour being roundly defeated in the May 1979 election. The door was left the open for the election of a Conservative government led by Margaret Thatcher and neoliberalism on steroids.

THE ASSAULT ON ORGANISED LABOUR: 1979 TO 2016

Margaret Thatcher and the Miners' Strike

For Labour, the adoption of monetarism and neoliberalism were the expedients of a government intent on bolstering British capitalism and unprepared to offer a radical alternative to the crisis. But for Margaret Thatcher these policies were both an ideological crusade and a strategy for restructuring British capitalism. She had been deeply influenced by right-wing economists such as Milton Friedman, the architect of monetarism (tightening the money supply as a way of dealing with inflation), and Friedrich von Hayek, who promoted a turbo-charged version of the market. Trade unions were seen as a thorn in the side of flexible labour markets in the neoliberal project and Thatcher launched a pervasive and vicious attack on organised labour on three fronts: driving up unemployment; attacking and defeating the miners who had twice humiliated the previous Conservative government; and introducing punitive and restrictive trade union legislation.

The world recession of 1981 was compounded by the catastrophic effects of monetarism driving interest rates up to 17 per cent. This precipitated a further decline in manufacturing jobs in Britain, a fall of 1.2 million between 1979 and 1981 that particularly hit industrial heartlands such as the West Midlands. By 1984 unemployment peaked at 12 per cent, a level unprecedented in the post-war period, severely weakening the ability of trade unions to resist and leaving workers vulnerable to an employers' offensive.

Another weapon in Margaret Thatcher's arsenal for trying to bring organised labour to heel were six pieces of legislation from 1980 onwards that amounted to draconian restrictions on the ability of unions to take lawful industrial action. The legislation restricted picketing and outlawed secondary action (sympathy action in support of other workers); introduced secret (and then postal) ballots for official industrial action; imposed seven days' notice that reduced unions' ability to respond quickly; and interfered with the running of unions' internal affairs and democratic processes. If there was any doubt as to the legality of a dispute (often on very technical grounds) employers could get injunctions from the High Court, who invariably ruled in their favour, so that if action went ahead unions could be charged with contempt of court, fined and potentially have their assets seized. Beyond attacking the legality of solidarity action and putting in place numerous obstacles to taking legal action, these measures were designed to pressure trade union leaderships to better police their members.

With the advantage of high unemployment and restrictive legislation Thatcher was armed for a direct confrontation with trade unions. She had learned the lessons of Edward Heath's calamitous Industrial Relations Act and a more comprehensive strategy for naked class war was articulated in the Ridley Plan[4] (Stubbs, 2014). The first element was to test the water and try to 'provoke a battle in a non-vulnerable industry, where we can win' (ibid.: 7). The defeat of the print workers and banning of unions at GCHQ[5] were fights to test their strength. Meanwhile the government made sure that they stockpiled coal to avoid the mistakes of Heath. A policy of 'cut[ting] off money to strikers and making the Union finance them' had the intention of starving the miners back to work (ibid.: 8). Finally, the government made plans to use the full force of the state to undermine the strike. In the words of journalist Seamus Milne (2014): 'Margaret Thatcher branded striking miners "the enemy within". The chilling catchphrase embodied her government's scorched earth onslaught on Britain's mining communities – and gave the green light for the entire state to treat the miners' union as outlaws.'

The 1984–5 Miners' Strike against pit closures was the longest and most bitter in British working-class history. From March 1984 until March 1985 they fought an unprecedented battle to defend jobs and communities against the full onslaught and might of the British state. The ultimate aim was to inflict a crushing blow to the trade union movement, which the Conservatives saw as a political threat to their plan for neoliberal reconstruction of the economy.

Whether a defeat was inevitable has been hotly debated and this decisive moment has been used by some to argue that the working class has little power and therefore trade unions have to look to the 'new realism' of partnership with employers. Despite the fact that the state threw everything at the dispute, it was the lack of will of trade union leaders that undermined the effectiveness of the strike. A severe setback was the TUC's failure to deliver its congress resolution in September 1984 to ensure the solidarity necessary from other unions to stop the delivery of coal and oil. Critical opportunities were missed for other groups of workers to fight alongside the miners and every time a second front looked like opening up, union leaders were quick to seize any offer rather than find common cause and widen the fight.

In contrast to the hand wringing of trade union leaders and their failure to deliver on their promises, there was a high level of international solidarity. And there were isolated but shining examples in Britain, for example no coal was moved by rail for 35 weeks in the Leicestershire coalfield and print workers at *The Sun* newspaper refused to handle a scurrilous front page likening the National Union of Mineworkers (NUM) leader Arthur Scargill to Hitler. A lifeline for the strike was provided by Miners' Support Groups that sprang up in nearly every town and workplace that were twinned with individual mines and villages. Despite huge levels of material support for the Miners' Strike, it still fell short of the systematic solidarity from workers that was necessary to win the dispute. By January 1985 the strike was in serious trouble as the drift back to work accelerated. An NUM special conference on 3 March voted narrowly for a return to work and at every pit the miners marched back to work behind their lodge banners. The NUM was badly wounded and the defeat was a severe blow for the morale of the whole working class.

The aftermath of the Miners' Strike was a brutal decade for workplace struggle characterised by a number of catastrophic defeats, unparalleled since the 1920s: of print workers in 1986 and workers at P&O ferries in 1988. A final blow was that after a brief respite against casualisation, the government abolished the National Dock Labour Scheme in 1989. The Transport & General Workers' Union's (TGWU) response was to become embroiled in a 'legal quagmire caused by balloting laws' and the outcome of its eventual unsuccessful strike was the dismissal of all its shop stewards at Tilbury Docks (Lyddon, 2020).

Despite the dark cloud of these defeats hanging over organised labour, some workers were still prepared to fight, even without the official support of their union. One of the highest-profile disputes in 1995 was when 500 Liverpool dockers were sacked and locked out for taking action in defence of workers from a smaller docks' employer. There was tremendous moral, financial and practical international support for the dockers and solidarity action included go-slows and boycotts in ports across the world of ships bound for and coming from Liverpool. Nevertheless, the 'left-wing' TGWU leadership's strict adherence to anti-union laws allowed employers to isolate and victimise trade union militants and return the ports to causal working. Paul Foot writes angrily in the foreword to *Solidarity on the Waterfront*:

> The leaders of one of the biggest and most powerful unions on earth [TGWU] sat back and watched the cream of their membership struggle alone. The leaders' excuse was that they were bound by the Tories' anti-union laws, under which the strike was technically secondary action. But the excuse was really a cover for the union leaders' own craven pusillanimity. This pusillanimity had grown with years of connivance with laws which were made to break their organisations. (Lavalette and Kennedy, 1996: viii)

The story of courage and tenacity on the part of the dockers and their families and the betrayal by trade union leaders is told vividly in the film *Dockers* (1990).[6] Although this dispute stretched into the period of the newly elected Blair government, no action was taken by that government to help this group of workers.

Excoriating and Threatening Strikers: New Labour, 1997 to 2010

The landslide victory of Tony Blair that installed him as prime minister in a New Labour government in 1997 was a reaction against Margaret Thatcher and subsequent Conservative governments. However, those that had voted against neoliberalism were to be disappointed. Blair had already set out his stall in 1995 when he won a controversial vote to amend Clause Four of the Labour Party constitution, ending the party's commitment to mass nationalisation and stripping away any veneer of socialism. The replacement clause that refers to the 'enterprise of the market and the rigour of competition' reflected Blair's commitment to continuing Thatcher's neoliberal policies. New Labour's friendship with and championing of big business were overtly trumpeted when, after three years, the then chancellor of the exchequer, Gordon Brown, boasted of Britain's lowest ever corporation tax for business and lowest ever capital gains tax for investors. References to ambition and aspiration became code words for justifying the amassing of great fortunes by individuals. Under New Labour the rich got richer. The increase in inequality produced by Thatcher, greater than that of any other Western country, was exacerbated even further under Blair. In 1997 the richest 1 per cent of Britain owned 17 per cent of the country's wealth. By 2006 under New Labour that figure had increased to 22 per cent (McIlroy and Daniels, 2010a).

Trade unions were wrong if they expected the shackles of the most restrictive legislation in Europe to be loosened; nearly every aspect of the previous Conservative government legislation remained intact. The rebranding of the party and government as New Labour and its mantra of modernisation and the 'third way' was intended to distance itself from its association with trade unions. Declarations of 'fairness not favours' and the promotion of social partnerships between trade unions and employers, to the benefit of the latter, underlined this decoupling from trade unions.

The Blair years saw a fall in the number of strikes and a low level of industrial struggle. The average number of strikes per million trade union members, having fallen to 32 in the 1990s, fell further to 19 for the period 2000 to 2009 (Brown, 2011). But despite trying to shackle trade unions to the interests of capital, the commercialisation and commodification of the public sector created constant and inbuilt tensions in workplaces, both in terms of emerging differential pay and working conditions and their erosion. Another source of tension was hostility to an ethos of the private sector foisted on workers that was antithetical to their motivation for working into the public sector. This was particularly true of teachers in the then National Unions of Teachers (NUT) who voted repeatedly against the use of testing in schools.

Sites of conflict at work continued to shift from traditional industries to public sector workers as the structure of the economy changed and teachers and lecturers in further and higher education and civil servants became more militant. Another locus of activity was in transport, communications and distribution. This was characterised by some unions being on a permanent war footing with a constant series of low-level skirmishes. For example, the fragmentation of the railways following privatisation resulted in widely varying rates of pay and terms and conditions. This embedded conflict was characterised by a cycle of regular strike ballots, threats of injunctions from employers and last-minute agreements.

Unofficial strikes did not disappear. The Royal Mail was deemed to be the most strike-prone group, with continuing high levels of unofficial strikes. Neither did the low level of struggle mean that there was an accommodation to Blairism, even in the upper echelons of trade unions. The Rail, Maritime and Transport Workers (RMT) and Fire Brigades Union (FBU)[7] broke their affiliation with the Labour Party and every other major union sent a shot over the bow by cutting funding to the Labour Party. Blair's continuation of Thatcherite policies alienated a layer of activists whose dissatisfaction contributed to voting in candidates to the left of their incumbents at the top of trade unions and saw the emergence of what was deemed the 'awkward squad' (see McIlroy and Daniels, 2010b: 150). Even though extended strikes did not disappear completely (for example the two-month strike of 5,000 nursery nurses in Glasgow in 2004), there was a shift to a pattern of one-day strikes accompanied by working to rule.

Renewing the Attack on Trade Unions: 2010 Onwards

In 2010 the Conservative Party led by Prime Minister David Cameron won the largest number of seats in parliament, but, falling short of a majority, formed an alliance with the Liberal Democrats under leader Nick Clegg. Far from the Liberal Democrats exerting any restraints on the worst excesses of the Conservative government, in the wake of the 2008 crisis they wholeheartedly supported austerity, which was the biggest attack on living standards and public sector services in post-Second World War history.

An opportunity for a significant confrontation with the government over public sector pension reforms was squandered. On 30 November 2011 2.5 million workers took one day of strike action. This was comparable to the number that struck against the Industrial Relations Act in 1971 and certainly the largest official strike since 1926. In total 29 TUC affiliates struck on the day, ranging from large public sector unions such as Unison, the NUT and the Public and Commercial Services Union (PCS) to a range of smaller public sector unions and professional associations. There were massive turnouts for demonstrations and rallies in big cities and smaller towns – and even the shires. Excitement and a sense of power turned to disbelief as the leadership of unions caved in with unseemly haste, under pressure from the coalition government, queuing up to sign the 'Heads of Agreement'. This detriment to pensions, especially for young people, meant substantially higher contributions and a pensionable age that increased well into the late sixties.

After the 2015 election, when they won a majority in parliament, the Tories introduced draconian legislation that limited the ability of trade unions to take legal action in what a *Guardian* newspaper editorial called a 'nasty and vindictive piece of legislation' (2016). The Trade Union Act of 2016 introduced a new requirement that 50 per cent of union members needed to vote in a ballot for industrial action. Particularly punitive was an additional clause for public sector workers (in health, school education, fire, transport and border control) that 40 per cent of those entitled to vote in a workplace had to vote in favour for strike action to be legal. Sidney and Beatrice Webb could have been writing today when they observed, 'It must not be imagined that either the ingenuity of lawyers or the prejudice of judges has been exhausted' (1920: 606). This legislation is open season for lawyers acting for employers to find a vast array of nit-picking irregularities and deficiencies that would get action declared illegal in the courts.

THE SHACKLES OF RESTRICTIVE LEGISLATION?

Some have argued that the movement of industrial relations into the theatre of the courts has been the most decisive factor in explaining the continued decrease in strikes in the new millennium. Trade union leaders, they argue, have absorbed the logic of legalism and are therefore 'unprepared to defy the law, and so utterly reluctant to provide any type of leadership' (O'Brien, 2018: 165), while workers feel more unsafe taking action because they can be easily and legally sacked.[8] But trade union leaders have never needed the excuse of legislation either to drag their heels in supporting action or to sabotage it completely. As Richard Hyman (1980) argues when he points to the collaborative tendencies of much nineteenth-century trade unionism in, for example, crafts, cotton, coal mining or iron and steel, there is a danger of viewing the past through rose-tinted spectacles. There have been periods when union officials opposed nearly all strikes, from 1940 to the mid-1950s for example. In the early 1970s, rather than leading strikes trade union leaders supported them because of pressure from below.

An exclusive focus on strikes as a measure of workers' combativity misses 'ballot victories' where workers have shown their preparedness to strike and employers have backed down. Ralph Darlington (2009) points to the 50 RMT ballots on the London Underground that took place between January 2000 and December 2008, with ballots leading to strikes on only 18 occasions. They were used as a form 'sabre rattling' to show the strength of feeling of workers and with no action taken managed to extract concessions. A longstanding Unite activist, steward and conference delegate quoted General Secretary Len McCluskey's words in his address to the 2018 Unite policy conference that specifically articulated such a strategy: 'in the last two years we have held more than three hundred industrial action ballots with over 90 per cent resulting in significant successes for our members without needing to activate the mandate'.[9] Even in the post-2008 period where average real wages have fallen, in some organised sections of the private sector large turnouts and massive majorities in favour of action have seen off attacks on pensions and secured above-inflation pay increases. A review of Unite's pay deal database in 2016 revealed that bargaining units in the aerospace, automotive, chemical and food sectors managed to get pay settlements that were either above inflation or in some cases double.

We should be cautious about simply treating balloting as a bureaucratic affair. It is almost impossible for the trade union bureaucracy to win both a high turnout and a decisive vote in favour on its own; such results are delivered by the massive involvement of rank-and-file activists. A Unite activist, Richard Morgan, writes about how they secured an 88 per cent turnout in a ballot:

> We used the industrial action ballot as an organising tool – going round the offices checking that everyone's address was correctly recorded by the union (at least 10 per cent of our members' details were incorrect), getting these details updated, initiating the ballot, going round again and checking that everyone had received a ballot paper, then going round again and checking that everyone had voted. Such activity meant that the workplace was taken into the postal ballot. (2016: 129)

Claims that restrictive legislation is throttling trade unions going on strike because leaders are intent on preserving the machinery of labour organisations and workers are intimidated by the fear of losing their jobs are important arguments to address as they fuel pessimism about struggle in the workplace. These debates will be revisited in Chapter 13 in the light of the case studies that form the backbone of this book. For now it is worth noting that the strikes and disputes in the time frame covered, 2015 to 2019, easily cleared legal hurdles. With regard to claims that workers are more reluctant to go on strike, the cleaners, care workers and council workers who took part in these strikes are some of the most vulnerable groups by dint of their low pay and often insecure contracts. Some had written these groups off and deemed them to be unorganisable, yet they scored significant victories in pay and working conditions. In 2018 and 2019 university staff – lecturers, researchers and support workers – lost 36 days' pay in their bitter fight over pensions.

Conservative governments have used legislation and all-out confrontation to try to contain and defeat organised labour, while Labour governments have preferred courting and co-opting trade union leaders on whom they have relied to police their own members. When workers move in a determined way, anti-union laws are not decisive – the most famous example being the defeat of the Industrial Relations Act in 1972. But even in the current period there are examples of workers simply ignoring the law, for example in 2018 when workers refused to cross the picket lines of the Glasgow women fighting for equal pay. This chapter has taken a broad brush to paint the contours of these struggles between trade unions, capital and the state in post-war Britain. However, key to understanding the microcosms of struggle that play out in the case study disputes covered later in the book is to open the black box of trade unions to look at the tensions and contradictions within them, which is the focus of Chapter 6.

6

Opening the 'Black Box' of Trade Unions

DEFEAT SNATCHED FROM THE JAWS OF VICTORY

In 2006 I had the bitter experience of having the rug pulled from under the industrial action in which I was taking part and having defeat snatched from the jaws of victory. Our university lecturers' trade unions[1] were operating a boycott of student exams and assessments in the spring and early summer of 2006 in support of our pay claim. Students were not going to be able to graduate and employers and the media were hysterical. Ordinary members of the unions were enthusiastic, serious and determined – they could see how effective the action was and sensed victory. But in June, after negotiations at the TUC's headquarters in London, the leadership of the newly formed University and College Union (UCU) suspended the action to ballot on a three-year pay deal with the employers' group. Without consulting members this powerful tool of action that had university managements on the run was jettisoned. The UCU's joint general secretaries announced that, 'we believe that this is the best that can be achieved within the current national negotiating environment' (Smithers, 2006). In my university hundreds of union members crammed into a lecture theatre with standing room only – for the largest meeting before or since – where they displayed fury and disbelief in equal measure that their collective power had been squandered. This anger turned to organisation and a couple of weeks later hundreds of activists met in London. The UCU Left was launched – an organisation that committed itself to building networks of activists and a democratic, accountable campaigning union that could work to mobilise and involve members in improving pay and conditions and defending progressive principles of education.

INSIDE OFFICIALDOM

There is nothing exceptional about the unfolding of events described above. Recognising the schism between a conservative, bureaucratic layer in trade unions and a militant rank and file has a long lineage. In their book *History of Trade Unionism*, first published in 1894, Beatrice and Sidney Webb point to a change in British trade unions in the second half of the nineteenth century when there was a shift in leadership, 'from the casual enthusiast and irresponsible agitator to a class of permanent salaried officers' (1920: 204, cited by Hallas, 1980). The way that the union bureaucracy, including national leaders and national, regional and local officials, occupies a social position that is different from the bulk of their members is captured vividly by Hal Draper when he writes: 'The effect ... is to isolate him from those he represents. He is removed from the discipline of the shop floor, from its dirt and dangers, from the immediate conflicts with the foreman and manager, from the fellowship of his workmates, to the very different environment of the office' (1978: 107).

As Darlington (2014a) points out, the behaviour of trade union officials has nothing to do with their individual personalities, competence or even politics; indeed many are recruited because of their effectiveness as fighters and leaders in the workplace. Trade union officials are not born as bureaucrats, but their behaviour is conditioned by and rooted in the very nature of their job. Union officials are paid by the union, not the employer, which means that their objective interest lies in perpetuating a situation where they intervene between labour and capital. Unlike representatives elected from among groups of workers, their interests diverge. This is put succinctly by Hal Draper:

> If the plant is closed, the official who negotiates the redundancies will not get the sack. Constantly closeted by management, he comes to see negotiation, compromise, the reconciliation of capital and labour as the very stuff of trade unionism. Struggle appears as a disruption of the bargaining process, a nuisance and inconvenience, which may threaten the accumulated funds of the union. Organisa-

> tion becomes an end in-itself threatening even the limited goal of improving the terms on which the worker is exploited. (1978: 107)

Therefore, rather aiming for an all-out victory, the role of officials is to 'get a deal done' in negotiations that take place within a prescribed framework. At high points of working-class confidence and action workers have insisted that negotiations are open. The Solidarity trade union in Poland in 1981 broadcast the negotiations to thousands of workers in the Gdańsk shipyard. During the Portuguese revolution in 1974, when thousands of workplaces were occupied and run by their employees, 118 workers insisted on going to the first meeting with management of the multinational firm Plessey. In 2012, during the Chicago teachers strike in the US, the negotiating team was expanded from single figures to 45 people.

The material benefits of union officials and their members diverge. The general secretaries of the 15 biggest unions in Britain earn enormous salaries hovering around the figure of £100,000 (plus additional benefits), and the layer of officials beneath them have secure jobs, sometimes with salaries that exceed those of the workers they represent. Preservation of the union's machinery, its finance and buildings becomes an end in itself. As we saw in Chapter 5, this is very well understood by governments, which is why legislation over the last 40 years has played on the fear that unlawful action could lead them into the courts, liable for damages of millions of pounds and the sequestration of funds. The highly centralised, hierarchical and bureaucratic nature of trade unions promotes a culture where all echelons of the machinery are kept under control. Financial resources and a monopoly of official communication are important weapons that constrain how far and enthusiastically policy is implemented.

The Labour Party was born from trade unions and socialist societies in 1900 as the Labour Representation Committee and trade unions retain strong institutional links to it.[2] But this historical alliance between trade unions and the Labour Party has often acted as a brake on strikes. Loyalty to the Labour Party, especially when they have been in office, has encouraged ministers to place pressure on union officials not to undermine (their) government or, when out of office, persuade them to drop action and strikes on the grounds that Labour might appear irresponsible – particularly in the run-up to elections.

Beyond Heroes and Villains

Some commentators have argued that 'final, fixed, situational, conflict bearing fissures' between union bureaucracies and their rank-and-file members are simplistic and unhelpful (McIlroy, 2014: 525). But to assert that there is a structural separation between paid officials and ordinary members and their representatives does not lead to a crude 'heroes and villains' approach to understanding the inner workings of labour organisations. Within trade unions themselves bureaucracies are not homogenous; there are tensions and contradictions within them. The existence of a union hierarchy means that there are significant differences between the general secretary who moves in circles of employers and other general secretaries and those union officials who are on the front line and closer to their members (Darlington, 2014a). Within trade unions, officials responsible for negotiating and full-timers dedicated to organisation and recruitment, who engage with ordinary workers as they find them, will have different motivations, experiences and influences. Neither are all union leaders the same. In the late 1960s and early 1970s many politically moderate national trade union leaders were replaced with new left-wing individuals – dubbed the 'awkward squad' – sometimes elected in response to greater industrial action and activity by workers.

The election of left-wing leaders can reflect members wanting a greater level of resistance and can strengthen the hand of union representatives and activists. But even though left-wing leaders are more likely to encourage workers' struggles, they are nevertheless subject to the inbuilt structural conservative and bureaucratic pressures that affect all union leaders with the consequence of holding back workers' struggles.

Shop stewards and representatives elected from among the people they work alongside form the backbone of trade unions, providing a link between the workplace and the national union. They are qualitatively different from trade union officials because they have a shared experience with their members, are networked in their sectors and have roots in and links with other activists in the community. The relationship between the scale and depth of union representation in factories, offices, warehouses, schools and hospitals and the level of struggle is symbiotic. A high level of industrial action will throw up new activists and workplace representatives and in times of downturn and low activity the opposite will be true. In the upturn in militancy in the late 1960s and 1970s strong independent workplace union organisation developed, often in the form of lay representatives that acted as an important counter-tendency to the bureaucratisation and accommodation of the official union leadership (Darlington, 2014a). Sometimes this distinction between the trade union bureaucracy and workplace representatives is muddied by the latter being given time off or even full-time release from their day jobs. This can lead to the incorporation of shop stewards into the official machinery, and senior stewards becoming relatively remote.

The Organising Model

One aspect of the tension in unions between keeping control of the machine and activism from below is captured by what is referred to as 'servicing' and 'organising' approaches. The servicing model relies on a professional trade union hierarchy that delivers collective and individual services to a passive membership, for example, at a national level, through bargaining and lobbying on questions of policy, and by providing case work support in the workplace. The organising model was imported from the US and particularly from the tactics used by the Service Employees International Union (SEIU) that focuses on building an active membership. There is no one agreed model, but in Britain Jane McAlevey's book *No Shortcuts: Organising for Power* (2016), about the experience of organising in the US, and her seminars have been highly influential and popular with many of the activists and union organisers interviewed for this book.

McAlevey's ideas have much to commend them. She talks about wresting power away from deal-making behind closed doors by expanding and involving the base of ordinary members who have never previously been involved in debate or activity. There are three strands to her argument. First, that organic leaders must be identified, that is, those who have the respect of and the ability to win over other workers. Second, McAlevey emphasises the importance of community in winning a dispute; as she puts it, 'organic ties to the broader community form the potential strategic wedge needed ... to leverage power' (ibid.: 28). The third strand is using structure tests to gauge how effectively an organic leader can win the majority on their shift or unit to taking 'high-risk action' by testing the ground with, for example, getting the majority of workers to sign a petition.

At its best McAlevey's approach codifies what committed activists have always done in the workplace instinctively, expressing and organising around the grievances of rank-and-file workers. As one activist turned organiser in the National Education Union (NEU) commented in 2019, 'it is not a work of genius ... it theorises what we have always done in [the] workplace'. One teacher activist argued that the importance of her approach was to shift members' mindsets to understanding 'that the union should not be doing for workers what they could do for themselves'. But McAlevey's organising model is in danger of tipping over into a prescriptive, linear and mechanical approach to struggle, particularly when it involves mapping workplaces by classifying members according to their propensity to become active. It misunderstands the volatile nature of disputes and how they can blow up in quick and unpredictable ways. Without doubt formal and informal 'structure tests' (for example indicative ballots) are an important mechanism for talking to members and making sure that activists are not isolated. But stipulating that a specific percentage of workers need to be won over to a particular cause can be used to stymie action by union officials and fails to recognise that a strike by a small group of workers may be spontaneous and can gather momentum to involve a much wider group. As we saw in Chapter 5, the Leeds women clothing workers and Pilkington's strikes in 1970 started with a few workers walking out, but tapped into wider grievances and snowballed into strikes that involved thousands of workers. Determining 'organic' leaders in advance misses the fact that during disputes new, unexpected activists emerge, sometimes overtaking existing leaders in their enthusiasm and militancy.

ALTERNATIVE STRATEGIES TO THE UNION BUREAUCRACY

In their frustration with the union bureaucracy activists have looked to a variety of strategies to overcome the inherent conservatism in the machinery of trade unions. This includes political strategies of building rank-and-file workers' organisations within unions, or other initiatives that aim to substitute for it such as breakaway 'red' unions, community unions or 'cyberunionism'.

Rank-and-File Movements

Ralph Darlington (2014b) provides a rich and thoroughgoing history of workers in trade unions in Britain organising independently of the bureaucracy to initiate and spread action, a history that can be traced back at least as far as the shop stewards' movement of the First World War. The Minority Movement of the 1920s, initiated by the Communist Party, was a high point in organising workers and recruiting them to a revolutionary agenda (Hinton and Hyman, 1975; Cliff and Gluckstein, 1986). There is insufficient space to cover this important history and there are significant differences between the politics and strategies of these early organisations and subsequent rank-and-file initiatives. Unofficial action is the outcome of workers taking action independently of the trade union machinery and, in the process, creating rank-and-file leaders. This was reflected in the activities of the International Socialists (that was to become the Socialist Workers Party) in the early 1970s. This organisation, with around 4,000 members, established 40 factory branches that spawned 16 sector-based newspapers – the most frequently produced being the *Dockworker*, *Rank and File Teacher* and *Carworker* (Callinicos, 1982).

In the post-war period and until the 1980s the Communist Party was an important left-wing influence in some sectors and trade unions. Their tradition of building 'broad left' coalitions inside the trade union movement assumes that the main division inside unions is between those with left- and right-wing ideas and that therefore the key task is to get left-wing candidates elected to trade union positions. In the 1960s and 1970s there was a powerful grouping in the engineering union that pulled together hundreds of shop stewards, activists and full-time officials. However, winning positions can shift attention from the involvement and action of ordinary members by assuming that a trade union bureaucracy with left-wing and progressive ideas can win disputes and concessions on behalf of workers. Although rank-and-file initiatives are uneven and eclectic, formal and informal, their fundamental premise is that workers in factories, schools, hospitals and warehouses are the agents and architects of struggle and the means of developing working-class consciousness. As we shall see, this has been borne out more recently by the networks formed on the ground to exchange information and experiences in fighting for health and safety in the Covid-19 pandemic in the face of the passivity of union leaders.

Breakaway Unions

As union activists and militants have despaired of breaking the grip of trade union bureaucrats and conservative officials, a parallel development has been the formation of breakaway radical trade unions. The most well-known example, borne out of the revolutionary syndicalism of the early twentieth century, is the Industrial Workers of the World (IWW) established in the US as early as 1905 and in the United Kingdom in 2013 (Heath, 2014). Eugene Debs, a founding member of the IWW, argues that, 'To talk about reforming these rotten graft infested unions which are dominated absolutely by the labour boss, is as vain and wasteful as to spray a cesspool with attar of roses' (Darlington, 2014a: 72).[3] It is a moot point whether the IWW was strictly a breakaway union; however, it did oppose the conservatism and elitism of craft unions by seeking to build an alternative mass industrial organisation.

Until the second half of the new millennium splinter trade unions in Britain were few and far between. However, two small organisations, the IWGB and the UVW founded in 2012 and 2014 respectively, have clocked up significant high-profile wins. The IWGB began as a breakaway from Unite and Unison when a group of cleaners became frustrated that their action was being undermined and their participation in union structures sabotaged in their fight for better working conditions at the University of London (a dispute covered in Chapter 9). Members of the UVW are mainly migrant cleaners and workers in other outsourced or low-wage industries and have strong associations with the Latin American community. Given the small size of UVW and IWGB and the financial and immigration status vulnerabilities of some of their members, their David and Goliath battles and victories on wages, working conditions and bringing outsourced workers back in-house have been stunning. Many activists will relate to the frustration with trade union bureaucracies that provokes activists to form breakaway unions, but isolating militants from supposedly less advanced workers is extremely problematic. Lenin, in his pamphlet *'Left Wing' Communism an Infantile Disorder*, grappled with this issue and took a hard line on those that advocated setting up breakaway trade unions. He is scathing about '*refusing to work within the trade unions*, on the pretext that they are "reactionary", and invent a brand-new, immaculate little "Workers' Union", which is guiltless of bourgeois-democratic prejudices and innocent of craft or narrow-minded craft-union sins'. He went on to argue that, 'If you want to help the "masses" and win the sympathy and support of the "masses", you should not fear difficulties, or pinpricks, chicanery, insults and persecution from the "leaders" ... but must absolutely work wherever the masses are to be found' (1921: 35)

Breakaway trade unions that assume a static situation where there is a division between militant workers and those that are deemed to be passive ignore the history of strikes and disputes where workers who have previously been inactive are propelled into action. For socialists, trade unions are a bridge to a wide layer of workers whose ideas are not fixed and predetermined but can be changed through experience and argument as part of a community with other workers. Being in a 'red bubble' avoids the necessity of persuading and winning over the majority of workers to support action and the outcome is to leave the majority of workers in the hands of old bureaucrats who then face a reduced challenge from below.

Community Unionism

Community unionism has entered the lexicon of trade union strategies in Britain in the new millennium. But there is no universal definition and the form and purpose of these types of organisation vary enormously within and between countries. The idea of community unionism is most developed in the US, often initially focused around worker centres. These centres grew up because of the poor record of some sections of the union movement in organising migrant workers and addressing their problems at work, as well as wider problems such as regularisation of immigration status and housing. Worker centres in the US combine servicing, legal help and organising, both for individuals and through collective campaigns. However, these centres have often relied on broad campaigns rather than putting workplace organisation at the centre of their politics. In her study Janice Fine notes, 'I was struck by how little workers' centres utilised the potential economic power of low wage immigrant workers themselves' (2006: 257). Some worker centres have moved from union-inspired radical pressure groups to operating as businesses – what is sometimes fashionably referred to as social entrepreneurship.

There is no real equivalent of worker centres in Britain, but there are examples of campaigns that carry some of the same politics. In 2001 the East London Communities Organisation (TELCO), with 40 affiliated community and union organisations, managed to persuade some employers to pay the 'living wage' as opposed to the minimum wage. This was followed by a campaign by the TGWU[4] in 2004, which recruited about 1,500 cleaners and got agreements with leading contractors in Canary Wharf in London, the site of global finance. Winning improvements in the pay and working conditions for one of the most badly treated and poorly paid groups of workers was a significant achievement. There are, however, limitations to some models of community organising which believe that a 'wide diversity of actors with a multiplicity of interests' can operate in the place of workers' self-organisation (Wills, 2008: 306). The issue of class is sidelined as 'workers' issues have been recast as community-wide concerns and class interests read through the lens of community, immigration, and race and religion' (ibid.: 309).

Since their inception trade unions have been rooted in communities from which they have drawn support and solidarity. This stretches back to the explosion of strikes in the East End of London in the 1880s that established the 'new unionism' of unskilled workers. Some have argued that this relationship between community and trade unions has been ruptured by industrial restructuring from the 1970s onwards. However, the groups that sprang up in every city, small town and workplace to support the strike of miners' in 1984–5 and twinned with communities at a distance shows that the idea of community is not geographically constrained. Cutting through extensive debates, the key question about community unionism is whether it supports struggles or whether broad-based alliances are a substitute for workers' action in the workplace. The strike by the Chicago Teachers' Union in the US in 2012 linked demands for improved pay and working conditions with the wider fight of working-class communities for a better education for their children. But it is important to underscore the fact that an all-out strike was at the heart of their victory. On the first day of the strike 35,000 teachers and their supporters marched through the heart of the city and subsequent days started with picketing their schools before attending mass rallies.

Cyberunionism

At the other extreme of emphasising geographical proximity some people have championed solidarity in virtual space – 'cyberunionism' – enabled by the growth of internet-based communication. It is argued that 'distributed discourses' that give a voice to those below enable rank-and-file activists to challenge the bureaucratic conservatism of trade union leaders (see Upchurch, 2014, for a discussion). An early example is the long-running Liverpool dockers' dispute (1995–9) which used the internet to create solidarity networks outside of Britain. However, a distinction needs to be drawn between workers using social media to build action and solidarity and 'slackivism' – the tendency by some to substitute passive, verbal and physical isolation to participating in real-time, real-space activity. It is the case that the monopoly of the established digital infrastructure of trade union bureaucracies allows them to control information and to circumvent democratic forums such as branches by using e-ballots. However, this does not guarantee them having a grip on a dispute, as was shown in the UCU pensions dispute. In 2018, after taking a round of well-supported strike action and in response to a poor deal negotiated by the leadership, messages from angry activists around the #nocapitulation hashtag spread like wildfire. Rank-and-file members were mobilised at short notice to physically lobby the National Executive Committee who were being asked to endorse the deal. As we shall see in later chapters, social media and internet-based technology have been important tools for enabling activists engaged in struggles but are not an alternative form of organisation or substitute for democratic trade union structures.

A DIALECTICAL RELATIONSHIP

To denounce, downplay or ignore the influence of officials is to absolve them of any responsibility for the task of attempting to encourage workers to fight back. There is a dynamic interplay at work where the official structures of the union offer the possibility of acting as a catalyst for increasing the potential scale of workers resistance. This in turn spurs a more favourable climate for the emergence of strong networks that can put further pressure on the bureaucracy. Union officials, particularly on the left, can play an important role in ways that advance rank-and-file confidence, initiative and organisation. Crudely counterposing, championing or vilifying the bureaucracy is an abstract and unhelpful formulation for socialists and activists in real-world disputes. Rather there is an ongoing process of working with the bureaucracy at the same time as having a critical relationship with it. This relationship of working both with and against the bureaucracy is summed up by the maxim of the Clyde Workers' Committee in 1915: 'We will support the officials just so long as they rightly represent the workers but will act independent immediately they misrepresent them' (cited in Briggs and Saville, 1971: 164). The following chapters go on to look at workers' struggles and how these tensions play out.

7

Striking Women: Still Hidden from History

LIFE-CHANGING VICTORIES

In 2018 and 2019 two images of women workers barely registered in the mainstream media. The first was in October 2018 when 8,000 angry women council workers in Glasgow took to the streets as part of a two-day strike to demand that the council delivered on their promise of equal pay. The soundtrack was a cacophony of klaxons, whistles, chanting and singing. Union banners were interspersed with homemade placards – one of which read 'Sell your labour – not your soul'. The demonstration was led by women and at the rally women who had never spoken before in public came to the microphone outraged that the Scottish National Party (SNP) minority council had failed to deliver on their promise of rectifying equal pay cases that had festered for over a decade. Since 2006, Labour-dominated councils in Glasgow had tried to circumvent equal pay by buying women off with paltry sums, setting up an arm's-length firm to impose inferior pay and conditions and then spending millions of pounds in court defending their flawed scheme rather than addressing the equality issues. These women won their fight for equal pay and in May 2019 their long overdue back pay started arriving.

The second image was a smaller, but equally noisy, celebratory demonstration of women care workers in Birmingham in May 2019 as they forced the council to back down from cutting their hours and wages. From 2017 onwards they had taken on their bosses who bullied them and tried to impose atrocious working practices, plunge them into poverty and dismantle their service. In 2019, after two years of backdoor tricks, threats and underhand tactics from employers, the care workers overwhelmingly won three ballots that translated into

82 days of strike action. This was the longest dispute of care workers in Britain. The demonstration was a jubilant celebration by women, young and old from diverse backgrounds, literally amplified by a megaphone. These momentous victories in Glasgow and in Birmingham were life changing, both in the material gains made by the women and in their transformation into activists and organisers.

This chapter begins by foregrounding these struggles in the multifaceted persistence of inequality in the workplace in Britain. The narrative goes on to place the action and campaigns of the women in Birmingham and Glasgow in the historical context of other important strikes of women. Some strikes are iconic in labour history, for example the Match Women's Strike of 1888 and the Ford Dagenham equal pay strike of 1968. However, other important struggles such as the Cradley Heath women chainmakers' strike in 1910, the Leeds Garment workers' strike (1970), the Trico equal pay strike (1976) and the Lee Jeans factory occupation in 1981 are under the radar and less celebrated in official trade union histories. This is due, at least in part, to the fact that they were struggles from below, and sometimes within the trade union itself. Yet these strikes and occupations are important for understanding how resistance and organisation can come from the least likely quarters and by the most oppressed and fragmented sections of the workforce.

THE PERSISTENCE OF INEQUALITY

Women are no longer marginal, but central to the labour force in Britain with their participation rate growing from 53 per cent in 1971 to 71 per cent in 2018 (ONS, 2019). In the late 1960s the women's liberation movement in Britain, part of the wave of civil rights movements that swept Europe and the US, made radical demands regarding the private and working lives of women. In response to mounting pressure from this movement and the aftermath of the Ford Dagenham women's equal pay dispute in 1968, the landmark Equal Pay Act (1970) and Sex Discrimination Act (1975) were passed. Together these pieces of legislation promised to address inequalities in the workplace and open up more and better opportunities for women. Five decades later the pay gap between men and women has narrowed but is stubbornly persistent. Discriminatory pay for women is alive and kicking. In 2019, hourly pay was 8.9 per cent less for women than men in full-time employment, and if part-time workers are included (who are relatively poorly paid) then this gap increases to 17.3 per cent (Francis-Devine and Pyper, 2020). Neither can women rely on the legal system to redress this injustice. Although an average of almost 29,000 equal pay complaints are received each year by employment tribunals, the vast majority (approximately 98 per cent) of recorded cases did not reach a full hearing (Webber, 2020).

In 2018, 40 per cent of women were working part time, often because it allows them to combine work and caring responsibilities, a burden increased by the decimation of the welfare state by austerity imposed in the wake of the 2008 financial crisis. Far from the introduction of 24-hour nurseries demanded by the women's movement in 1971, childcare costs for young children in Britain are exorbitant: the highest in Europe except for Switzerland. Most of what is on offer for small children is in the private sector, and with the average cost of a full-time nursery place for a child under two running at £242 a week (£329 in London) it is way beyond the means of all but better-paid women. However, working part time is not always a choice: since 2008 there has a big increase in the proportion of women who are part time because they are unable to find full-time work. Linked to this is the rise in the number of women who say that they are underemployed and would like to work more hours than their current job gives them (TUC, 2015).

Since 2008 there has been spectacular growth in the number of women who are self-employed, with the number increasing from 1.05 million to 1.45 million by the end of 2014 (ibid.). However, far from the emancipatory language of 'entrepreneurialism' and 'being your own boss', this condemns many women to poor earnings. Two in five women with a self-employed income earned less than £10,000 a year and on average women earned 30 per cent less than men from self-employment (ibid.). Importantly, self-employed women miss out on benefits, such as statutory sick pay and statutory maternity pay, denying them basic entitlements from the welfare state. Women comprise a disproportionate number of workers on temporary contracts or ZHCs, predominantly in the retail, health and social care sectors. However, these contracts are not confined to low-paid work but are endemic in further and higher education where growing numbers of women have little employment security and few opportunities for progression. The power of the employer not to renew a contract exposes women to increased risk of discrimination in relation to pregnancy and maternity.

The fallout from the economic crisis of 2008 has threatened to arrest or reverse some of the hard-fought gains for women in the workplace. Women, who make up two-thirds of the public sector workforce, have been disproportionately affected by austerity and cuts in public spending that resulted in pay freezes, job losses and the intensification of work. Between 2008 and 2014 the biggest reduction in the number of women employees was in retail and manufacturing; in financial services women accounted for three-quarters of the net job losses in the sector (TUC, 2015). Half the net job growth of women's work since 2010 has been in low-paying sectors, with the biggest growth in jobs being in health and social care.

A harsher climate at work after the 2008 crisis has made it harder to bargain for equality. Flexible working has been more difficult to secure, and managers' attitudes towards working parents and carers has got tougher. New rights to promote sex equality, such as the right to request flexible working and the new system of shared parental leave, will bring benefits to some women, but mainly those in permanent employment with partners also in secure work. For many women – in insecure work, receiving low pay and working part time – these rights are much more difficult to access. For example, by 2018 the government reported that a miniscule 2 per cent of eligible parents took up shared parental leave (Walker, 2018).[1] Low-paid workers, even if eligible, could not afford to reduce their incomes further. Despite legislation in the Equality Act 2010 that is supposed to protect workers from sexual harassment in the workplace, a survey by Focus on Labour Exploitation (2021) revealed a significant gap in its enforcement. It reported that 33 per cent of, mostly women, cleaners had suffered behaviour ranging from sexualised comments to sexual assault.

However, despite the toll of austerity since 2008, trade union density for women has been fairly constant at just under 30 per cent while for men it has declined. Given the detrimental trend in women's employment, with part-time work, reduced public sector employment and increasing temporary work, women's membership has been fairly resilient. But union membership among younger women has fallen significantly from an already low base, at least in part explained by their higher likelihood of temporary contracts and zero-hours working.

The Covid-19 pandemic placed a disproportionate economic and emotional burden on women. Their concentration in the health and care sectors put them on the front line where they were perilously endangered by the lack of PPE. Overrepresented as workers in health and social care services, those from BAME backgrounds were even more lethally exposed (Fawcett Society, 2020). This is reflected in the stark statistic from the Unison trade union that 72 per cent of all health and social care staff who died in the first few months of the pandemic were black (Unison, 2020). Even when working from home, women were doing the lion's share of childcare in the first lockdown in March 2020. Mothers were more likely than fathers to have moved out of paid work and juggling work alongside paid work put an immense strain on their well-being and mental health. Bringing together the effects of gender and ethnicity during the Covid-19 crisis means that BAME women, in particular, have suffered disproportionate physical, psychological and financial impacts (Fawcett Society, 2020).

But women have not been passive in the face of the Covid-19 crisis. The threat posed to the health and safety of teachers and their communities from prematurely opening schools during the pandemic brought thousands of young women teachers into activism as union representatives. As we shall see, the victories achieved in Glasgow and Birmingham laid the groundwork for fighting for health and safety in the workplaces of their members. The next section highlights some significant historical strikes by women, often in the most difficult circumstances, against low pay, precarious work and discrimination.

FROM THE BIRTH OF CAPITALISM TO THE GREAT UNREST: 1888 TO 1914

Organised struggles by women stretch back in history. An editorial in *The Times* in 1808, describing unrest among Lancashire weavers, said that: 'The women are, if possible, more turbulent and more mischievous than the men. Their insolence to the soldiers and special constables is intolerable' (Lewenhak, 1977: 22). In the late nineteenth century trade union leader Ben Tillett described the Match Women's Strike and victory in 1888 as 'the beginning of the social convulsion' (Raw, 2011: 222) which produced the movement for new unionism among unskilled workers. Rather than being 'helpless waifs' or tragic victims at the mercy of the market these very young women, working in horrendous conditions, defied assumptions about gender to take action against a powerful employer. While this episode is well embedded in popular history, 20 years later the less well-known strike of women chainmakers in the Black Country, north-west of Birmingham, fed into the mood that saw trade union membership increase dramatically between 1910 and 1914, after a wave of industrial unrest swept across the country in 1910 and 1911.

During the nineteenth century the Black Country, and Cradley Heath in particular, became the centre for chain-making in Britain, where smaller chains were hand worked by women and children in small, cramped forges or outbuildings next to their homes. This was literally 'sweated labour' – hot, physically demanding and poorly paid. The women were isolated and unorganised and middlemen creamed off profits from the gap between the market price of chains and what they paid domestic workers who had to individually negotiate a price. The average male wage at the time was 27 shillings per week (£70 per year) while before the strike women workers were earning five to six shillings for working a 54-hour week – one-fifth of the average male wage (Barnsley, 2010).

In 1910 employers cajoled 1,000 women into signing a form opting out of the newly introduced minimum wage of the Chain Trade Board – (literally) banking on the fact that many were illiterate and had no idea what they were signing. In August 1910, when the women, led by Mary Macarthur, founder of the National Federation of Women Workers (NFWW), demanded that the new minimum wage be implemented, the employers immediately reacted by locking the women out. Mass meetings and the new medium of cinema were used to bring the situation of the striking women to a wider audience. They gained immense popular support and the nearly £4,000 raised by the end of the dispute made it possible to pay the women strike pay of five shillings per week – the equivalent of what they earned for a 54-hour week. Barnsley quotes the County advertiser for Staffordshire and Worcestershire who commented: 'The women's blood is up, and they mean to have their emancipation day ... It is inconceivable that the employers can hold out against such a demand ... The chain workers are at last learning the secret of united action' (ibid.: 44).

The strike lasted for ten weeks and the women held out until the last employer agreed to their demands. The victory of the women chainmakers not only made them significantly financially better off, it also boosted their confidence. The membership of the NFWW in Cradley Heath grew from 400 before the strike to 1,700 at the end of it – a significant number in a small town.

This strike was part of a wider movement of struggles against 'sweated labour' that swept the country. Boston describes how, 'Like a chain reaction, in the hot summer of 1911 women in London, jam and pickle workers, rag-pickers and biscuit-makers, bottle washers, tin-box makers, cocoa makers, distillery workers – all sweated factory workers earning between 5s and 10s. a week – came out on strike for higher wages' (2015: 69).

The female membership of all trade unions more than doubled from 166,803 in 1906 to 357,956 by 1914. Textile unions in 1914 accounted for two-thirds of organised women, while the other third was mainly drawn from the general distributive and clerical unions. The growth of the suffrage movement and the explosion of trade unions among unskilled workers in 1880s also reignited the demand for equal pay first raised in 1834 by the Grand National Consolidated Trades Union's journal *The Pioneer* (Frow and Frow, 1989). In 1888, Clementina Black, an English writer, feminist and pioneering trade unionist closely connected with Marxist and Fabian socialists, moved the first equal pay resolution at the TUC. However, this was a shot across the bows of what was to be a protracted fight for equality at work that would not be put on the statute books for another eight decades.

BREAKING THROUGH: 1945 TO 1970

Both the First and Second World Wars provided a fillip for women in the workplace. Fixed ideas about women's role in the home and their ability to do only particular work were quickly jettisoned. The number of women in engineering rose from 97,000 in 1939 to 602,000 in 1945 as labour shortages meant that they were given access to jobs that were previously closed to them, and thousands of nurseries were opened to provide childcare. When it quickly became clear to women during both wars that they were being employed on lower wages they demanded equal pay. A successful tram and tube strike in 1918 resulted in the government establishing a special enquiry as to whether there should be wage equality between men and women across all industries. However, during the Depression of the 1930s, in some quarters, women workers were blamed for unemployment among men and many trade unions consequently became hostile to the issue. During the Second World War the momentum for equal pay was resumed with the Equal Pay Campaign Committee established in 1941. In 1943 there were extensive strikes by men and women in the engineering and aircraft factories of Scotland and the North East, whose demands included equal pay.

A government report was issued in October 1946 and the Equal Pay Campaign Committee organised publicity and public meetings to support its findings. However, the 1950s was a bleak decade for women; post-Second World War politicians were not inclined to concede equal rights at work to women. The Beveridge Report that laid the groundwork for the welfare state envisaged a traditional role for women. In a racist and xenophobic clause it advocated that 'mothers have vital work in ensuring the adequate continuation of the British Race and of British ideals in the world' (Beveridge, 1942: 53). Attempts to move women back to the home after their increased participation in the workforce were made easier by the TUC's acquiescence to the Labour government's closure of nurseries. The TUC's 1948 annual report demonstrated complete ideological harmony with Beveridge when it claimed that, 'There is little doubt in the minds of the General Council that the home is one of the most important spheres for a woman worker' (Davis, 2020).

The Equal Pay Committee Campaign was persistent.[2] In 1955, after mass public campaigning including demonstrations and petitions, a scheme established equal rates of pay for men and women doing the same job in the non-industrial civil service (through seven equal annual instalments from 1955 to 1961). Similar agreements were made for white-collar women in local government, the NHS, nationalised industries and teachers and nurses. By 1964 the Labour Party manifesto called for a charter for the rights of all employees to include 'the right to equal pay for equal work'; in 1965 the TUC conference followed with a similar resolution calling for the equal treatment of, and opportunity for, women in industry.

If the government and TUC had wanted to kick the issue of equal pay into the long grass they were thwarted by two factors. By the late 1960s growing demands for a more liberal society were epitomised by progressive legislation such as the Abortion and Sexual Offences Acts of 1967 and the Divorce Reform Act of 1969. The National Joint Action Campaign for Women's Equal Rights organised a huge demonstration in London in 1969. Equal pay was a key demand of the embryonic women's liberation movement, but it was the strike by women sewing machinists in the Dagenham Ford plant that moved the issue of equal pay to centre stage.

On 29 May 1968, 187 women machinists at Ford's River Plant in Dagenham, Essex, walked out of their factory in protest against sex discrimination in grading. The strike, which lasted for three weeks and brought Ford's entire production line to a standstill, was resolved when Ford asked Barbara Castle, Labour secretary of state for employment, to intervene and persuade the women to return to work. However, some of the activists considered that the dispute was not the success claimed by union officials. With a 7 per cent pay rise their percentage of the men's grade increased from 85 to 92 per cent, but equal pay was not on the agenda. In 1978 one of the shop stewards, Lil O'Callaghan, reflected that: 'We mucked it up. We should have left it open to fight another battle on another day' (Moss, 2015: 43). Another shop steward echoed this disillusionment arguing that, 'although we did get more money, we did not gain the point, we won the battle but lost the war' (ibid.: 44).

The notion that the strike was an unequivocal success was further undermined by the fact that the sewing machinists had to wait until 1984 to have their work regraded after a five-week strike in November and December that year. However, although the women did not initially gain the regrading that they had demanded, the 1968 strike was seen as a landmark in British industrial relations, widely credited with prompting the 1970 Equal Pay Act. It had wider political implications – the action of the Ford women and the discussion about equal pay both reflected the burgeoning women's movement in Britain and was a further impetus to its growth.

FROM THE MILITANT 1970S TO THE 1980S DOWNTURN

Three strikes and occupations, important in this period, are largely hidden from history. The militancy of the women workers, their independent action and the pressure they put on their trade unions do not always sit comfortably with official labour histories. The upsurge in militancy in the early 1970s is often associated with men working on building sites, in mines and on the docks, yet it was a huge strike by women garment workers in Leeds that opened this explosive chapter in struggles from below. The Ford Dagenham strike has an iconic status in labour movement history, but the Trico strike for equal pay in 1976 is more significant in terms of its duration, the militancy of the women workers and the complete victory they achieved. In 1981 women workers at Lee Jeans faced the closure of their factory and the loss of their livelihoods. However, despite the global recession, the onslaught against the working class by the Thatcher government and the withdrawal of support from their union, they occupied their factory and saved their jobs.

Leeds – United!

In February 1970, clothing workers in Leeds began to walk out of their factories on unofficial strike in support of their demand for a shilling an hour pay increase. The action snowballed as strikers marched around the city, from factory to factory, calling on others to join them until 20,000 workers were on strike. The strike was an angry response to an agreement between the National Union of Tailors and Garment Workers (NUTGW) and the Clothing Manufacturers Federation which perpetuated low wages, left differentials between men and women's pay unchallenged and foisted the first nationally negotiated productivity agreement on them. The relatively inexperienced strikers, unburdened by the structures of union conventions, went from one workplace to another to bring other workers out on strike. This powerful weapon of what came to be known as 'flying pickets' was later used in the struggles of the building workers and miners. The enthusiasm for and participation in the strike by thousands of women and the battle of activists with union officials and bosses is captured in a film, *Leeds – United!*

After four weeks, full-time trade union officials proposed a return to work so that the union could negotiate with employers. The rank-and-file committee was outmanoeuvred and the shop stewards voted to recommend a return to work by 83 to 76. But a mass meeting of 4,000 people the following day unanimously voted to stay out. A further meeting of stewards was put under intense pressure by union officials, and subsequently agreed once again to recommend a return to work. This time the strikers went back. The calling off of the strike was seen as a betrayal by activists; the strike was still spreading and with more determined leadership could have won their demands in full. For the women involved in the strike there was a qualitative change in their understanding of the potential of the trade union movement taking strike action and a recognition of the role that they could play.

The Trico Equal Pay Strike

Although many assumed that the Equal Pay Act of 1970 would secure equal pay as a right and be the 'crowning glory' of women's campaigns, the Trico case demonstrates that there were many battles ahead. In their account of the struggle, *Trico: A Victory to Remember*, Sally Groves and Vernon Merritt look at how this 21-week strike for equal pay was the most significant struggle for women's rights arising from the expectations created by the Equal Pay Act. The five-year gap between the passing of the bill (1970) and its implementation (1975) gave employers plenty of time to reorganise the labour process to circumvent the spirit of the Act. The tribunal that reviewed cases brought under the Act turned out to be a labyrinth of complexity and showed a strong pro-employer bias.

Trico was a factory that produced windscreen wipers in North London. As in many workplaces, segregation on the basis of gender was a major source of cost savings to employers. Women worked the day shift and men exclusively worked the night shift that attracted pay that was a third higher.[3] The tinder for the dispute was when the company downsized and redeployed five men onto the same line as women workers. The company justified the greater reward to men because of their alleged 'flexibility' – a classic avoidance tactic on the part of firms – and said they had no intention of implementing equal pay. On 24 May 1976, in the face of intransigence by management, 400 women workers at Trico walked out, including 98 who were not even in the union; they demanded that their employer pay equal wages for equal work. When the strike was made official by the Amalgamated Union of Engineering Workers they were joined by 150 men, mainly skilled toolmakers. Shockingly, the vote for all-out strike action in support of the women by other male workers was defeated by 60 to 40 per cent.

Management were not prepared for the tenacity of the women strikers and thought the dispute would be short-lived. Eventually the employers decided to take the dispute to an industrial tribunal (since 1998 called employment tribunals). The refusal of the union and strikers to attend, because of the poor record and pro-employer bias of such tribunals, was vindicated when the tribunal ruled in the employer's favour. However, although the employer had won the tribunal, in the face of the persistence of the strike and with growing solidarity from the labour movement, they had no strategy for ending the dispute. In October 1976 the company caved in and conceded to all the demands, most importantly a common operational rate regardless of sex that gave the women an average increase of about 20 per cent. It is worth quoting the words of Monica Harvey, one of the strikers, describing how strike action rather than relying on employment tribunals and the solidarity of other workers won the dispute:

> Our equal pay strike is over. We went back to work on Monday having won a complete victory. What a bloody mockery this makes of the decision of the equal pay tribunal. They said we had no case but we've shown them otherwise. We refused to be pressurised by the decision and our refusal to go back to work until we won showed that strength and spirit is what really matters. A strong shop floor organisation is the way to win. It is worth a million Equal Pay Acts. We now have the same piece rate as the men. Together with a £2.30 wage increase, this gives us a total rise of £8.70. The financial support given to us by workers in factories and workplaces all over the country was great in keeping us going. This support amounted to hundreds and thousands of pounds, all given by brothers and sisters who committed themselves to our victory. Our full-time officials have been great … [they] refused to give even the time of day to the various pathetic offers made by management during the strike. (Groves and Merritt, 2018: 162–3)

It was the first time that a trade union had boycotted a tribunal and then gone on to negotiate a settlement despite the tribunal's decision ruling against them. In the face of women workers having to confront American-style picket-busting convoys of lorries and scab labour, a legal system stacked against them and the majority of their male co-workers working, this was a stunning victory and the longest successful strike for equal pay in British history.

The Lee Jeans Factory Occupation

As we saw in Chapter 4, the Conservative government, elected in 1979 with Margaret Thatcher as prime minister, proceeded to launch a concerted and all-out assault on the organised working class by driving up unemployment and introducing restrictive trade union legislation. This attack on organised labour, coupled with a slump in the global economy that disproportionally hit manufacturing, made this a difficult and demoralising time for the British working class. Yet, in parallel to this gloomy story, between February 1981 and March 1982 there were three successful factory occupations across Central Scotland in workplaces with predominantly women workers: Lee Jeans in Greenock, Lovable Brassiere in Cumbernauld and Plessey Capacitators in Bathgate. Garment production is notoriously geographically footloose and has a history of firms taking the money (government subsidies) and running (relocating elsewhere) when they end.

Women workers at Lee Jeans walked out in support of shop steward Margaret Robertson who was sacked after she complained about the freezing conditions in the factory (see Figure 7.1). The response of the management was to propose closing the Greenock Lee Jeans factory in February 1981 and relocating it to Northern Ireland where the subsidies had not dried up (Robertson and Clark, 2019). Margaret Robertson talked about how the strike began:

> We got the workers together and had a meeting in the canteen … And I put it to them, if they [management] said no to everything we offered, we could stay in the factory and not leave. And it was put to the vote, the workers agreed and the plan was set … I went into the meeting and was told that they had rejected our proposals

> and the factory would close as planned. And the next thing we were barricaded in, and we occupied the factory. (Ibid.: 341)

Their union – the NUTGW – was supportive to begin with but officials dragged their heels in making the strike official, and then quickly got cold feet claiming that they could be fined for breaking the trade union laws. However, the spinelessness of the NUTGW was in stark contrast to the support the women got from other workers, with shop stewards from the local shipyards and miners voting to put a levy on their wages to support the Lee Jeans workers.

The 24-hour occupation lasted for seven months until a management buyout of the factory eventually vindicated the workers' argument. This struggle is important because occupations are a particularly labour-intensive form of action and relatively uncommon in Britain – successful examples are even rarer (Sherry, 2010). The occupation was particularly unusual because it took place at a low ebb of class struggle and represents a successful example of action against a factory closure by a young female workforce resisting the relocation of global capital.

Figure 7.1 Lee Jeans workers occupying their factory in 1981: 'Send out for 240 fish suppers!'

Source: John Sturrock.

From 2010 onwards two disputes involving low-paid women working for local councils in Birmingham and Glasgow simmered slowly and then erupted into lively campaigns and strikes. The two-day strike of, mainly women, council workers in Glasgow to get the equal pay they had been promised in October 2018, and the 82 days of strike action by women care workers in Birmingham against the council's proposal to cut their hours, are landmark struggles of women workers.

BIRMINGHAM CARE WORKERS: STRUGGLING TO DEFEND CONTRACTS AND PAY

The Birmingham care workers' dispute is part of a long story of how women's work has been consistently undervalued. Looking after small children, the elderly and those with health needs is seen as an extension of what they do in the family which provides a convenient justification for their low pay. There is little acknowledgement of the complex skills expected of care workers that include high-level 'people skills' as well as the ability to administer drugs, check pressure sores and use hoists and other techniques to move people. Their work involves supporting stroke victims and those undergoing drug detoxification and providing palliative care, and as such they are expected to get National Vocational Qualifications (NVQs) in these skills (at level three, an NVQ is equivalent to an A-level). 'Caring' for people in their homes evokes a cosy narrative, but the reality is that significant safety issues affect care workers, such as having to walk or drive in unsafe places in the evening or night. Women care workers report being carjacked, robbed and assaulted in the course of their work. During the Covid-19 epidemic these care workers were invisible to the government – despite them being on the front line and in the high-risk category as they cared for vulnerable people. Their health and safety was seriously compromised by the abject failure to supply PPE.

A War of Attrition

In 2017 there were 460 Home Care Enablement Workers, nearly exclusively women, employed by Birmingham City Council to care for people when they were discharged from hospital. Although 95 per cent of the women were in the Unison trade union they had no history of or experience in taking strike action. In fact, Unison had previously treated them as a special case on the basis of their jobs to exempt them from taking action in previous disputes. These workers had already had their pay cut by up to £5,500 in 2011 when the then Conservative City Council took away the additional pay for weekend working. The second attack on working conditions and wages was in April 2017 (under a Labour-run council) when employers tried to introduce split-shift contracts. This would have meant working 14 hours a day for eight hours' pay because of having to hang round between shifts and clients.

A ballot for strike action in November 2017 was won with a 54 per cent turnout and 99 per cent vote to strike and the council backed down. But later the same month the council came back with a new tactic to wear down the women by embarking on a protracted period of interviews for voluntary redundancies – often without union representation. The proposed new shift patterns were used as a way of bullying women into taking redundancy: one worker had night blindness and that was an excuse for managers suggesting that she was not 'up to the job'. The number of care workers was reduced by a further 48 per cent, leaving just 270 staff. The process stalled a new vote for action as trade union legislation makes it almost impossible for unions to ballot while voluntary redundancies are being negotiated. Undeterred, the response of the workers was three two-hour strikes in January, February and March 2018. Caroline Johnson, a Unison organiser, explains how management tried to pull the rug from under them:

> The strike action was called off after management appeared to make a concession by allowing our care workers to organise their own rosters. This was no easy task in the light of the 'give and take' necessary to make sure that the needs of their clients were met, and at first the employers appeared to respond positively. However, far from this being taken seriously by management, it turned out that this was a further delaying tactic on their part to string us along until after the local elections in May 2018 and to make sure that the ballot for action ran out of time.

The next move by Birmingham City Council in June 2018 was to commission consultants, Newton Europe, at an eye-watering cost of £12 million to review care and elderly services. In the worst kind of scientific management these consultants proposed reducing all contracts to part time: 14, 21 or 22.75 hours a week. Caroline explained:

> They even dictated how many workers they wanted in each category: 88 people working 22.75 hours; 79 working 21 hours and so on. As 60 per cent of workers were on contracts of 30 hours or more this represented a serious wage cut in a job that was already badly paid. These consultants also wanted to introduce fixed rotas that changed on alternate weeks meaning that women would be unable to take the second jobs that they needed to make ends meet. We reballoted for an escalation of action and won a 73 per cent yes-vote.

Organising on Three Fronts

The organising strategy was inspired by Jane McAlevey's book, *No Shortcuts: Organizing for Power*. Rather than workers being ciphers of slick top-down campaigns by the national union, Caroline explained how the women were put at the centre of the action: fully informed and involved in making decisions about tactics. They were able to build on the strong base of a high density of union membership – the legacy of the success of unions winning significant gains in an equal pay dispute in 2008. Nevertheless, this paper membership had to be galvanised into 'agents of their own success' and in practical terms this meant painstakingly building a vote for strike action among women who had previously been passive members of the union. A huge amount of work and systematic effort was put into getting the vote out – initially with telephone calls made to all of the 460 workers. Mandy Buckley, a new activist, explained:

> We kept a live spreadsheet of members to record who had been contacted and their responses. Many of the women did not use social media and we encouraged them to come into the union offices to set up email addresses to make sure we had quick and easy contact to counteract rumours, whispers and 'fake news' from management.

> We had regular mass members' meetings to increase participation and everything was voted on.

There were three parallel strands to the campaign, the first of which was strike action. By the end of the dispute the women had been on strike for 82 days starting with token two-hour strikes that escalated to five-day strikes (see Figure 7.2). These were bolstered by rallies and demonstrations supported by other Unison branches and trade unions. The second strand of the campaign was to get public support and union solidarity. Beyond frequent stalls held in the city centre of Birmingham the women targeted the wards of Labour members of the inner cabinet of Birmingham City Council who had proposed the cuts. As Mandy explained, 'Thirty of us would turn up armed with leaflets, we gave them out at supermarkets and went door to door ... they [Labour councillors] hated it.' Solidarity from branches of Unison and other unions was important in terms of generous donations that topped up strike pay and sustained morale. Unison nationally paid out £180,000 in strike pay, but an equal amount was raised through the donations of other trade unions and individuals.

Figure 7.2 Birmingham women carers lobbying on the first day of their strikes June 2018
Source: Geoff Dexter, *Socialist Worker*.

The third strand of the campaign, in the face of a war of attrition by management over a long period of time, was to keep the momentum of the dispute going; the branch did this by raising the political awareness of the women, and placing their dispute in the context of wider politics and historical struggles. Film screenings were held of *Made in Dagenham* about the Ford equal pay strike and *Pride* that told the story of a gay miners' support group from London twinned with a traditional community in South Wales during the Miners' Strike of 1984–5. The branch organised a trip to London to lobby members of parliament and visit the headquarters of their union, Unison. Women brought their daughters, many of whom had never left Birmingham before. In September 2018 the women lobbied the Labour Party conference, delegations went to the anti-Trump demonstration in July 2018 and the Stand Up to Racism conference, and they led the protests at the Conservative Party conference.

In May 2019, in the face of the determination and tenacity of the women, Birmingham City Council withdrew the proposals in their entirety. The women had defended their jobs and livelihoods. Even before the draconian changes to wages and contracts, these women were at the forefront of austerity. In 2018 they earned £9.81 an hour (plus a night allowance), but their take-home pay in real terms was not even as much as before the financial crisis of 2008. Many women faced real financial hardship and even meeting the most basic needs meant remortgaging their homes, depending on overdrafts and getting into debt. Contrast this with the position of those working for the consulting company, Newton Europe, which was paid £12 million by Birmingham City Council to make the 'efficiency costs'. Their lives could not be more different. The average pay for consultants was £47,000, while the post-tax profit of the firm in 2017 was £9.7 million on a turnover of £41 million.

In 2018 one of their graduate employees described the perks: 'Newton funds weekends for things such as skiing, adventure sports, camping and the Edinburgh Fringe. We get generous expenses so daily restaurant dinners are paid for as well as nice hotels. We have great summer and Christmas parties as well as fortnightly activity evenings, dinners and nights out paid for by the company.'[4] As one of the care workers said, 'they are living in a different world' – while the daily expenses of consultants are paid for, some care workers were going to supermarkets late at night to find food at reduced prices. But in a much more modest way the final settlement of their dispute changed the world of the care workers. Not only had they saved their jobs but two teams had taken responsibility for organising their own rostering which transformed the work–life balance of some of the women. For the first time Gloria was able to negotiate the same shifts as her partner so they were not just 'ships in the night' and Cherry was able to vary her hours so as not to be permanently on weekends – she had 'never cooked a Sunday dinner for her family'.

The victory of these women in May 2019 was only a brief respite from struggle. As the Covid-19 pandemic erupted in March 2020 they were once again on the front line. Unable to 'work from home' on Zoom they were supporting vulnerable people in their homes. Two teams directly supported those that had been discharged from hospital after having the virus. They drew on the combativeness and confidence they had gained from their dispute to fight for the PPE they so urgently needed.

FROM THE COURTS TO THE STREETS: THE EQUAL PAY STRIKE THAT SHOOK GLASGOW

A Festering Sore

To understand the way that unequal pay was entrenched in Glasgow we have to go back to the new collective agreement, signed in 1997 between local authorities and trade unions. For the first time such an agreement was aimed at harmonising the conditions of employment of 1.5 million manual and white-collar workers in local government in the United Kingdom. The rollout of these Single Status Agreements was partly an attempt to stem the tide of equal pay claims in employment tribunals. For decades the division of labour in local councils meant that work predominantly done by women, such as cleaning, catering and clerical work, although graded the same as jobs done mainly by men, were deemed unsuitable for productivity bonuses and other schemes based on output. As a result large discrepancies in pay resulted from access to overtime and bonus schemes, from which gravediggers, refuse collectors and road sweepers (mainly men) were more easily able to benefit. The outcome was that workers in these jobs could potentially take home thousands of pounds of additional income annually.

In the early days the main unions, Unison, Unite (and its predecessor the TGWU) and the GMB, representing the women council workers in Glasgow did not cover themselves in glory. At worst they signed up to pay structures that were clearly discriminatory and entrenched unequal pay, and at best they showed a lack of leadership and dragged their heels in fighting these cases for equal pay. However, as we shall see, reinvigorated branches and new activists from Unison and GMB played a crucial role in the success of the strike along with new, energetic and committed union organisers.

In Glasgow the conflict over the pay and reward system had stretched over more than a decade. One interviewee explained that:

> In 2005 Glasgow's Labour City Council tried to protect themselves from an avalanche of equal pay claims by bussing workers into leisure centres just before Christmas, where they felt they had no alternative but to sign a piece of paper that accepted a settlement that had no regard for length of service and was far below what they were entitled to legally. Many women signed but a few hundred held out.

In 2009 a second tactic by the council for circumventing equal pay and driving down pay and conditions was the setting up of an arm's-length organisation – Cordia – for some jobs and services. In January 2015, in an attack on the contracts of carers similar to that mounted by Birmingham City Council, the employers proposed a three-shift system that meant in effect being at work for 14 hours for only eight hours' pay, with time to kill between clients.

A Perfect Storm

In Glasgow the issue of equal pay was a running sore. Cases taken to court in 2011 and 2015 were long and convoluted, and resulted in rulings against claimants. However, three elements came together to make a perfect storm which triggered and inspired an equal pay campaign in September 2017. First, two rulings of the Court of Session (the highest civil court in Scotland) in May 2017 and August 2017 ruled against the exclusion of women from bonuses, because Glasgow

City Council had not demonstrated that their job evaluation scheme was fit for purpose. Second, the political composition of Glasgow City Council changed. Labour had been the dominant party in the council since 1945 and had governed uninterrupted since 1980, but in May 2017 the SNP won minority control of the council with 39 out of 85 seats. After much lobbying by the trade unions before the elections the SNP had promised to resolve these longstanding equal pay issues in their manifesto. Third was the significant victory of a lawyer, Stefan Cross, and his firm, Action4Equality, with a long history in fighting for equal pay. At an employment tribunal in 1997 he won £1 million for 2,000 school meals staff (mainly women) in Cleveland, North East England. In 2005 in Glasgow he recruited and represented a group of women who had refused to settle with the council. After Cross won settlements for this first wave of women, thousands of others signed up as clients in a second wave of claimants. The Facebook group they formed cascaded information to other women who then demanded that their unions fight for equal pay.

Buoyed by the decision of the court in May and August 2017, stewards, activists and organisers in Unison started a campaign the following September that involved and was led by the members. An apparent victory occurred in January 2018 when the SNP minority council voted not to appeal the decisions of the Court of Session and agreed to begin negotiations to settle the equal pay cases. Unison and the GMB increased pressure on the council and ramped up the momentum of the campaign with a demonstration held on 10 February 2018, when thousands of women marched in suffragette costumes with placards that read 'Equal Pay or We Walk Away'. Whether through incompetence or deliberately dragging their heels, Glasgow City Council failed to agree a timescale and plan for settling the equal pay claims. The anger of the women who had been waiting years to get justice boiled over: 200 women with claims submitted had already died waiting for their cases to be resolved. A consultative ballot in May 2018 showed resounding support for action.

'Strike, Strike, Strike'

The GMB and Unison organisers describe the huge pressure that they were under from below as their women members called for action to force the council to settle their claims. The Unison organiser describes how at one meeting she was met with a packed room of angry women chanting 'strike, strike, strike' and demanding that they were kept 'strike ready'. In September 2018 the ballot for strike action showed the depth of support for action as Unison and GMB ballots of home carers, cleaners and caterers in Cordia returned a 99 per cent 'yes' vote. These hugely impressive votes did not simply happen; support was built through a system of stewards in workplaces and through thousands of phonecalls made by activists at the headquarters of Unison and the GMB.

The build-up to the two-day strike in October was met with huge enthusiasm by the women. Both Unison and the GMB opened up strike headquarters as a space for activists to organise. Photographs of the strike plastered an entire wall in the Unison headquarters that was open and busy round the clock before the strike. The GMB took over a deserted bank next to their offices: three months after the strike it was still filled with placards, and with whiteboards with handwritten rosters for making phone calls and picket duty. The participation of women who had never been on strike before, and their self-organisation from below, was remarkable. They organised their own picket lines by phoning up members and forming WhatsApp groups in different areas of the city to ensure that all workplaces were covered and members could be based in their communities.

The mood among the women was that they wanted action and were not interested in previous union turf wars. In Unison there were regular campaign meetings and a comprehensive shop stewards structure built from previous struggles that meant they could coordinate action across the seven job families affected by the dispute (including cleaners, caterers, care workers, social workers and educational support workers). Despite the illegality of secondary picketing, refuse workers supported the strike by refusing to cross picket lines and lost two days' wages. The GMB organiser told me that on the day of the strike she had an angry phone call from the shop steward in the parks and recreation department demanding that pickets were put in place so that they could support the strike by not crossing picket lines.

The strike brought the city to a standstill and rocked the council (see Figure 7.3). They were forced to promise to make good on the equal pay claims. In February 2019 the strikers held a celebratory rally

Figure 7.3 Glasgow women strike for equal pay
Source: Andrew McGowan.

outside Glasgow City Chambers as Glasgow City Council ratified a package of payments worth £548 million to compensate approximately 16,000 equal pay claims. In May 2019 thousands of low-paid women celebrated again when they received details of their equal pay settlements. One worker said the offer left her weeping, feeling 'over the moon' and ready to book a family holiday to Jamaica. Unison steward Lyn-Marie recounted: 'My members were delighted with the settlements … but we haven't won the lottery, we got what we had worked hard for and deserved … It's not just the settlement that has changed people, but the experience of fighting back.' The victory of the Glasgow women has given confidence to other women. By 2021 the battle for equal pay had spilled into the retail sector where women working in the stores of the giant supermarket Asda were demanding parity with, mainly male, workers employed in distribution centres.

LESSONS, LINKAGES AND LEGACIES

In Glasgow and in Birmingham the working lives of the women who took part in these campaigns and strikes were precarious. Their low wages meant that securing the basics for day-to-day living was a struggle: successive rounds of hikes in energy prices were met with dread, and many of the women had one or even two additional jobs to make ends meet. Juggling demanding work and caring commitments in the family was a constant challenge. However, rather than the togetherness of the factory or office, cleaners and care workers can go without regular contact with their co-workers and narratives of isolation have been used to suggest that collective action is difficult or even impossible. The GMB organiser in Glasgow talked about having to overcome the union bureaucracy's pessimism about the possibility of organising these women workers and to confront negative attitudes of 'we can't find them, they won't come to meetings'. On the contrary, in Birmingham and Glasgow anger at injustice, and systematic organisation by union organisers and grassroots members, brought women in their hundreds to meetings in packed rooms.

Solidarity was critical for winning these disputes. In Birmingham pressure on employers from coordinated industrial action in other branches of Unison could have won the dispute more quickly. However, support from other workers in terms of invitations to speak at meetings and donations to the strike fund were crucial in supporting the women materially, raising their morale and keeping the dispute going. In Glasgow ordinary union members from other departments simply defied anti-trade union legislation, refused to cross picket lines and closed down the city. The giant trade union, Unison, the only player in Birmingham and one of the key players in Glasgow, threw significant resources and support behind both disputes. As we shall see later, there are instances where Unison has tried to exert much stronger top-down control over the pace and rhythm of some disputes in comparison with others. However, in Glasgow union officials and organisers from Unison and the GMB were forced to respond to members' demands for action that welled up from below. In the case of Birmingham, a combination of constant pressure exerted by the branch on the national leadership of Unison and a supportive regional secretary allowed the action to be led by initiatives from, and the democratic decisions of, the branch. However, the self-limiting action of Unison nationally and their failure to 'sing the victory of these workers from the rooftops', as one activist put it, meant the opportunity to generalise the lessons of the dispute were squandered.

Beyond the crucial material benefits won, the disputes threw up new leaders and activists and engendered a surge in confidence. In Birmingham the Unison union branch was invigorated, with new stewards elected and undergoing training. Mandy, who became a leading activist among Birmingham care workers, talked about how the dispute was 'life changing', and how she had gained in confidence in negotiating with management and supporting other workers. From someone who had never spoken in public before, she addressed meetings and large rallies and conferences all over the country. Similarly, Lyn-Marie, Unison convenor for Parklands, Property and Land Services in Glasgow described how:

> The dispute gave the branch a super boost ... we increased membership, we felt believed in ... we became leaders and incentivised others. I would never have believed that I could speak in public ... but after speaking in George Square at the demonstration I thought I could do anything ... I had no script ... I told it like it is.

A woman shop steward representing educational support workers in Glasgow talked about the way that the strike had empowered her:

> I had been working abroad and had good jobs ... when my marriage failed I came back with two children in 2006 and had to start at the bottom ... I am a support for learning worker ... but I felt browbeaten and worthless ... this strike has meant coming alive again ... feeling empowered ... they can't ignore us anymore.

Full-time Unison organiser Jennifer McCarey sums up how coming out of the 'legal theatre' and onto the streets, waging the battle on the picket lines rather than in the courts, empowered the women: 'Collective action builds a more confident and combative layer in every workplace ... we inspired each other, we raised the bar. The strikers became something different ... they became campaigners, fighters.'

The strike was a huge victory and is a landmark dispute. For the women the sums of money will be life changing. The victories of these workers are hugely important and continue the tradition of women's struggles. Even though these disputes are beacons of hope to other workers they have not had the attention they deserve. Their enthusiasm, tenacity and determination is inspirational and shows how groups deemed to be 'vulnerable' can fight and win. Chapters 8 and 9 continue this theme by looking at how migrant and precarious workers have also shredded preconceptions about their inability to organise, and have achieved stunning successes in getting better working conditions and wages.

8

Migrant Workers: Here to Stay, Here to Fight

MIGRANT WORKERS AND BRITISH CAPITALISM

Consider these two scenarios. The first is in Ireland in December 2005 when 100,000 Irish, Polish, Lithuanian and Latvian workers demonstrated together against attempts by Irish Ferries to employ migrant workers on worse pay and conditions than Irish workers. The second is in Britain in January 2009 when hundreds of construction workers took unofficial strike action under the slogan of 'British Jobs for British Workers' in protest against the employment of workers from the EU on lower wages. The former reflects the possibility of solidarity between indigenous and migrant workers and the latter a divisive situation in which fearful workers turn on 'foreigners'. Migrant workers are not passive victims of capital and neither are they unorganisable: they have often been at the forefront of strikes, union organisation and political activity in Britain and the US. This chapter looks at how migrant workers have become an integral part of the British working class, which they have both transformed and been transformed by. It looks at the key struggles in the workplace that shaped this process after 1945 and how the solidarity of workers is not automatic but has been fought for by trade union activists and socialists.

A Reserve Army of Labour

The British referendum vote to leave the EU (Brexit) in June 2016, and the debate before it, unleashed a torrent of xenophobia from right-wing populists. Not only did the Labour Party fail to counteract their poisonous narratives, but they directly contributed to a hostile climate for migrant workers. In 2012 Ed Miliband, the then leader of the Labour Party, claimed that when in power his party 'had got it wrong' in 2004 by allowing uncontrolled immigration from the newly joined states of Central and Eastern Europe (BBC, 2012). Jack Straw, former Labour home secretary, went even further, calling it a 'spectacular mistake' (Philipson, 2013). Yet despite migrants being recast as a problem by all mainstream political parties since the 2008 crisis, these workers are central to the functioning of British capitalism by working in the least skilled and worst-paid jobs. At its most naked, witness the desperate measures to fly in workers from Romania in April 2020 during the Covid-19 pandemic to pick salad and fruit that would otherwise have rotted in the fields of Britain.

The share of foreign-born people in total employment in Britain increased from 7.2 per cent in 1993 to 18 per cent in 2019, before falling to 16 per cent in September 2020 when some migrant workers returned home as a result of the Covid-19 pandemic (Migration Observatory, 2021). Where foreign-born workers are employed varies greatly across occupations and sectors, though the growth in their share of employment has been fastest in lower-skilled jobs. In 2002 there was only one low-skilled occupation (food preparation) in the list of the top-ten occupations with the highest share of foreign-born workers (Rienzo, 2015). By 2014 there were at least five low-skilled occupations on this list. In 2014 foreign-born workers comprised 43 per cent of workers in elementary plant processes (industry cleaning, packing, bottling), 33 per cent in cleaning and housekeeping and 32 per cent of process operatives (food and drink processes) (ibid.). Crudely put, recent migrant workers filled labour shortages on the bottom rung of the labour market.

Migrant workers play a distinct role in capitalism both as a 'reserve army of labour' and as a way of increasing profits by enabling bosses to raise the rate of exploitation. Migration has always been high on the agenda of governments and employers, particularly in developed capitalist economies such as Britain. The use of migrant workers conveniently allows the receiver country to externalise the costs of renewing the labour force. Because migrant workers are generally young, healthy and educated elsewhere, the state uses them to fill gaps in the labour market without them being a 'burden on the state'. One worker from the Solidarity trade union I interviewed in Poland expressed his resentment at what he thought was a colonial relationship in which

Poland paid for the welfare and education of young workers who were then poached by other countries, Britain in particular.

The State as Manager of Migration

The uneven and dynamic nature of capitalism produces a constant tendency for workers to move to seek better opportunities as production is closed down in some places and opened up in others. More jobs and better wages in more developed countries (and within countries) are a pull to workers who are 'pushed' from places where jobs have been lost or wages are low. Capitalists need the constant movement of workers, but they also need to retain skills to compete with other capitalists. Governments have often been preoccupied with preventing emigration and the loss of skilled workers. Therefore, states play a central and active role in managing outward and inward flows of labour across their boundaries and creating a hierarchy of legal entitlements.

Britain regularly and systematically poaches workers such as nurses, teachers and social workers from developing countries, often leaving them with shortages of vital workers. After eight former communist countries joined the EU in 2004[1] employers in existing EU states were able to cherry-pick workers from Central and Eastern Europe, leaving those countries with shortages of some skills, in health care in particular. The loss of doctors from Romania meant that some small towns could no longer perform operations because of a loss of anaesthetists to Western Europe (Hardy et al., 2016). In 2009 one worker from Poland told me how a bus company from the Midlands hired a hotel in Warsaw and put leaflets around the city's bus depot. Large numbers of drivers who turned up to the meeting were promised what appeared to be good wages (in comparison with the £1 an hour they earned in Poland), and the following week he and 20 drivers moved to the UK. The company paid the minimum wage, but the scant contract he showed to me stipulated that there were no set hours and the drivers would have to work as needed. This meant that some weeks they did not make the salary they had been promised because of a shortage of work and other weeks they could be called up in the middle of the night and worked 60 hours. When they complained to the managers, an English test was introduced with no notice as a way of disciplining the drivers: three drivers were sacked, and the remainder left and walked straight into one of the many jobs available elsewhere.

But the arrival of migrant workers is not a constant upward trajectory. They come to work. Some settle and become part of the British working class, but others return home when the jobs dry up. The cataclysmic impact of the Covid-19 epidemic on work in the hospitality, manufacturing and retail sectors has led to a mass exodus with an estimated 1.3 million foreign-born workers leaving Britain between March and December 2020 (O'Connor and Portes, 2021).

THE 1950S TO THE MID-1970S: UGLY RACISM AND THE BEGINNINGS OF RESISTANCE

Britain has always been a multiracial society. Since Roman times there has been an African presence. As the possessor of the largest empire in history, its shameful role in trading slaves and with London as the financial capital of the world, the movement and exchange of people was inevitable. For 500 years there has been a black community in London and in the 1850s Liverpool had the largest Chinese community outside of China. The denigration and dehumanisation of other nationalities and races was necessary to justify Britain's role in the slave trade and its inhumane treatment of black people. Ideologically it underpinned Britain's imperialist expansion into India and Africa, and Ireland before them. Racism and xenophobia seeped into every corner of British society. Successive waves of migrants – Huguenots, German émigrés and Jews – have all suffered derogatory stereotyping and prejudice. After the colonisation of Ireland in the seventeenth century the Irish faced systematic discrimination. Irish workers, driven out of Ireland by the famine in the 1840s and extensively employed in the frenzy of railway construction in the 1840s and 1850s, were a target for hostility. Denvir describes several days of riots around Penrith (north-east England) on the building of the Lancaster to Carlisle railway when 'rioting was commenced by English navvies who were unwilling to allow the Irish to work on the same part of the railway' (1894: 158). Such was the frequency and violence of attacks in this period that he notes that it read like the 'record of a nation in a state of civil war' (ibid.: 159). Therefore, racism and xenophobia integral to its role as an imperial power was endemic to Britain, and, as we shall see, persisted in the post-war period.

'Colour Bars' and Complacency

The docking of the *Empire Windrush* from Jamaica in 1948 heralded a new period of migration to Britain. Acute labour shortages in the wake of the Second World War drove the search for new sources of labour to rebuild the British economy. Active recruitment for industrial labour led to the Labour government's British Nationality Act in 1948 that granted British citizenship to people from colonies and ex-colonies. By April 1956, eager to fill labour shortages, London Transport and the Hotel and Restaurant Association had sent representatives to Barbados offering jobs and loans to pay for the passage to England. The Conservative health minister, Enoch Powell, actively recruited cleaners and nurses for the newly formed NHS.

In many quarters the new migrants from the Caribbean, India and Pakistan met with open hostility, reflecting prejudice that had been honed over centuries. There was no law against discrimination and a 'colour bar' operated all over the country in clubs, pubs and restaurants. In the workplace new migrants were concentrated in low-skilled and badly paid work with few opportunities to progress. They faced blatant discrimination by employers; trade unions hostile to black workers often colluded with management to entrench segregation and discrimination and some local trade union branches operated 'colour bans' or 'colour quotas'. In 1955 bus drivers in Wolverhampton banned overtime 'as a protest against the increasing numbers of coloured workers employed', and the TGWU insisted that no more than 52 of the city's 900 workers should be black (Fryer, 1984: 376). In West Bromwich bus workers staged one-day strikes in 1955 against the employment of a solitary Indian conductor.

A groundbreaking response to racism came from an unexpected quarter. After their conference in 1947, the Musicians' Union resolved to oppose the colour bar wherever it appeared. After a sustained campaign they got the colour bar lifted in Mecca ballrooms in Nottingham, Birmingham, Streatham and Sheffield in October 1958. In August 1958, in response to a white riot lasting several days that terrorised the black community in Notting Hill, the Musicians' Union quickly produced a statement condemning racial discrimination; signed by 27 entertainment industry celebrities, it appeared on the front page of *Melody Maker* on 5 September (Blackwell, 2017). One outcome was the formation of the Stars Campaign for Interracial Friendship in 1959 that organised many events and opened interracial clubs. It is hard to comprehend how brave this was in the poisonous climate of racism at a time intensified by the active presence of fascist organisations such as the White Defence League and Oswald Mosely's British Union of Fascists.

However, this positive approach was absent from the leadership of the trade union movement who exhibited contradictory behaviour. On the one hand the leaders of the TUC urged branches to welcome and aid overseas workers, and on the other pleaded with the government for immigration controls to 'maintain full employment' and to stop the market being 'flooded with cheap foreign labour and a serious deterrent to trade union bargaining power' (Ramdin, 2017: 199). Despite exhortations from some TUC delegates for concrete action to combat discrimination, from 1958 to 1964 no policy initiatives were forthcoming and the TUC failed to acknowledge racial injustice. In 1965 an even more negative tone appeared when the TUC General Council announced that 'immigrants lacking an adequate knowledge of English and English customs are presenting growing problems' (ibid.: 199). Although several unions were critical of this stance, casting immigrants as a 'problem' enabled the TUC to justify and support the radically restrictive proposals in the Labour government's White Paper on Immigration in 1966.

Combatting Racism on the Ground

But it is far from the case that all trade unionists were racist. Beryl Radin (1966) points to examples of socialists and trade unions combatting racism and discrimination, welcoming migrant workers and actively trying to recruit them. In her fascinating account based on interviews in different British cities, trade unions and workplaces in the early to mid-1960s, she documents very different responses on the ground from regional officers, branch officials and rank-and-file members. These included organising leaflets in the native language of prospective members, producing booklets welcoming migrants and encouraging them to join a trade union and holding Saturday morning surgeries to hear and help with problems faced by immigrant communities. The Smethwick Trades Council disaffiliated from the local Labour club because of its ban on interracial members and were involved in the formation of a local interracial organisation (ibid., 1966). Rather than a uniform picture of white racist workers and trade unions, Radin points to the important microdynamics of the workplace, noting that quotas on black workers on the buses were sometimes administered on a citywide basis or sometimes even by one depot or garage. In other factories and offices anti-racists and socialists argued and organised against racism and for solidarity with migrant workers from both inside and outside the workplace. However, in the climate of the time this was not always easy. Roger Cox, an engineering steward in North London at the time, recounts how workers in another factory, ENV, supported striking Asian workers at the Woolf factory in 1965:

> You have to understand that the level of racism against immigrants was terrible, far worse than today. Raising support for an Asian strike meant having an argument – even in a factory with a lot of Irish workers, many of whom had come to Britain after the war. Nevertheless, ENV adopted the strikers and organised a levy of all workers. (Prasad, 2017: 76)

But migrant workers were neither passive victims of racism nor were they reliant on sympathetic white workers to win battles on their behalf. Active opposition to discrimination by black migrant workers was demonstrated in the April 1963 Bristol bus boycott when West Indian activists exposed the Bristol Omnibus Company's long-standing colour bar against black crews (Dresser, 1986). Inspired by the politics of Martin Luther King and the American civil rights movement the activists announced a boycott on 29 April 1963. Black activists and the black community, with the support of the local MP Tony Benn, students at Bristol University, local Christian groups and the Campaign for Nuclear Disarmament, secured a significant victory against racism in the workplace when a mass meeting of 500 TGWU members on the buses voted to end the colour bar on 28 August.

Asian Workers Move into Struggle

In the mid-1960s Asian migrant workers moved into confrontation with employers, often having to battle with racist union officials as well as bosses (Mahamdallie, 2007). One important organisation that confronted workplace discrimination was the Indian Workers' Association (IWA): established in Coventry as early as 1938 it became a national organisation in 1958 (Gill, 2013). It actively struggled for unionisation as a way of encouraging the integration of Punjabi migrants into working-class life 'as part of a mission of building a class-based movement' (ibid.: 558). The IWA played a critical role in getting rid of the corrupt broker system that entrenched segregation by privileging white workers. Middlemen acted as intermediaries between the foremen and workers who they controlled through systems of patronage that involved allocating jobs, higher piece rates and overtime. From 1965 a wave of strikes, assisted by the IWA, attacked this racketeering system. Increasing levels of unionisation lessened the need for intermediaries, but it was the increasing number of Indian shop stewards that dealt a lasting blow to the broker system (ibid.). Therefore, the emergence of Indian shop stewards had a significant impact on shaping the struggle and on labour relations in the West Midlands.

The bitter defeat of a strike of Asian workers at Imperial Typewriters in Leicester in 1974 is a salutary lesson in the potential for divisiveness. In the words of Yuri Prasad, the fight 'involved a struggle against a vicious multinational firm, a racist local union and the fascist National Front, which organised some white workers at the factory under the guise of the White Workers of Imperial Typewriters' (2017: 79).

A speed-up in one section of the factory, where all the workers were Asian, triggered a walk out by 50 workers that resulted in a 50 per cent fall in production. The TGWU that represented them failed to bring out white workers and refused to make the strike official. This was a signal that management could sack the workers who initiated the walkout with impunity. Despite picket lines being regularly attacked by the police and fascists, local factories and the community threw their support behind the strike. But this was not enough to win a strike where the TGWU failed to recognise racial divisions and preferred to perpetuate a divisive situation rather than build a united front comprising all workers. They paid a heavy price when Litton Industries decided to close down Imperial Typewriters and both black and white workers lost their jobs.

1976 ONWARDS: THE GRUNWICK STRIKE AND ANTI-NAZI LEAGUE TURN THE TIDE

The growing threat of the fascist National Front and the culmination of decades of struggle by workers of South East Asian origin started to shift trade union members and their leaders away from complacency towards, or even collaboration with, racism in the workplace. But two developments in the mid-1970s proved to be turning points in the relations between trade unions and black and migrant workers. The first was a strike in 1976 by a group of Asian workers, mainly women, who walked out of the small Grunwick photo-processing factory in Willesden, north-west London. The tenacity and courage they showed in a strike that lasted until 1978 transformed the politics of race in the labour movement and challenged ideas about passive migrant women workers. The second development was the formation of the Anti-Nazi League (ANL) in late 1977. One of the largest social movements in the post-war period, it mobilised the support of a wide range of workers, musicians and celebrities across the political spectrum to successfully challenge the rise of the fascist National Front and embed anti-racism in the labour movement.

'We Are Those Lions, Mr Manager'

Yuri Prasad captures the importance of the Grunwick dispute:

> Grunwick was to see some of the greatest scenes of black and white working class solidarity in the history of the trade union movement – with mostly white workers travelling from all across Britain to join a group of overwhelmingly Asian women outside their factory gates; pickets of up to 20,000 people taking on mounted police and the paramilitary-style Special Patrol Group; mass arrests and brutal beatings meted out to pickets; the use of police agent provocateurs to discredit the strikers; and post office workers refusing to touch scab mail, despite threats of exclusion from leaders of their own union, Grunwick was all this and more. (2017: 84–5)

The Grunwick factory employed 440 people, mainly women of East Asian descent, who were expected to labour under harsh conditions. They faced bullying, constant threats of dismissal, strict surveillance, suffocating heat and the indignity of being forced to put their hands up if they wanted to go to the toilet (Anitha and Pearson, 2018). The walkout in August 1976 was triggered by punitive targets and the bullying behaviour of management insisting on overtime at short notice. Jayaben Desai, the leader and iconic image of the strike, demanded her 'cards' and insisted she was leaving and not being sacked, famously saying: 'What you are running here is not a factory, it is a zoo. Some are monkeys who dance on your fingertips. Others are lions that can bite off your head. We are those lions Mr Manager' (Ramdin, 2017: 285).

The strikers joined the APEX (Association of Professional, Executive, Clerical and Computer Staff) trade union who immediately made the strike official and provided strike pay. But it quickly became apparent that there were two dynamics at work. One was the rhetorical and tepid support of the TUC and APEX who, in a bid to take the heat out of the struggle, pushed for the dispute to go to the Advisory, Conciliation and Arbitration Service. The other dynamic was the momentum of the enthusiastic and tangible solidarity from other parts of the trade union movement, feminists and anti-racists. Pivotal to the dispute was the decision of the local branch of Union of Post Office Workers (UPW) at Cricklewood to stop delivering mail coming in and out of the factory which effectively crippled the operation of the business. The local strike committee called for mass pickets that began in June 1977 and lasted until November 1977. The largest picket was on 11 July 1977 when an estimated 18,000 people turned up representing the whole spectrum of the labour movement: miners, hospital workers, teachers, car workers and dockers (see Figure 8.1).

Support from the official trade union movement was a welcome change from hostility and the lack of support to which they had been subjected in the previous decade and it raised expectations of success. But the TUC and other union leaders were under pressure from the Labour Party to play their role in controlling explosive high-profile struggles on the streets and in the workplace. Specifically, their anxiety to defuse the Grunwick dispute was driven by the overarch-

Figure 8.1 On strike at Grunwicks
Source: Socialist Worker archive.

ing desire of the trade union bureaucracy to deliver and protect the Social Contract – an austerity plan agreed between the unions and the Labour government (see Chapter 5). The rug was pulled out from under the dispute. The Cricklewood post workers, who had boycotted Grunwick's, were laid off by their management and the UPW leadership ordered them to end the action. They held out for a few more weeks and then narrowly voted to go back to work.

The APEX union insisted on an end to mass picketing with the threat of removing strike pay. In November 1977, with the virtual abandonment of the strike by the leadership of APEX and the TUC, the Grunwick strikers went on hunger strike outside the TUC demanding more and effective support. APEX responded by suspending the hunger strikers and withdrawing their strike pay. Unable to effect any change, the strikers abandoned their action after 670 days on 14 July 1978 without any of their demands being met. Jayaben Desai was bitter about their shabby treatment: 'Official action from the TUC is like honey on your elbow. You can smell it, you can see it, but you can never taste it' (Prasad, 2017: 89). Yet the effects of the strike reverberated well beyond the picket line. Asian workers, now part of organised labour in Britain, transformed themselves and changed the labour movement whose ranks they had joined. The reason for the defeat of the strike lies in the vacillating and supine behaviour of the trade union bureaucracy which wound down and abandoned a dispute that teachers, miners, dockers and hospital workers had supported in their thousands.

The ANL

The ANL, formed in late 1977, grew rapidly to become the National Front's most memorable opponent, garnering huge popular support. It recruited between 40,000 and 50,000 members, distributed over five million leaflets and sold around one million anti-Front badges and stickers. The mass opposition that the ANL mobilised in workplaces, on the streets and through its festivals managed to smear the National Front with the noxious label of Nazi (Copsey, 2017). The strength and high profile of the ANL came from the wide spectrum of support reflected in its steering committee that included: four Labour MPs; a representative of the Searchlight anti-fascist organisation; Miriam Karlin, an actor of British Jewish descent; and two members of the Socialist Workers Party. The range of endorsements was huge and varied, with sponsors including 40 Labour MPs, two football managers and numerous writers and actors. Support for the ANL from young people was bolstered by the appearance of high-profile bands at a march and massive carnival in London in 1978 with an audience of up to 100,000 people. Its sister organisation, Rock Against Racism, organised events up and down the country.

A plethora of other groups sprang up under its umbrella – Students against the Nazis, Football Fans against the Nazis, Skateboarders against the Nazis – with the aim of challenging racist and fascist ideas. So successful was the ANL in mobilising abhorrence to racism and celebrating multiculturalism that in 1978 nearly every teacher in the

North London school in which I taught, including all the senior staff, went en masse to a central London rally organised by Teachers against the Nazis. Workers radicalised by the struggles of the early 1970s and the increased activities of Asian members were won to active anti-racism and participated in their thousands in ANL-led events.

In the general election of 1979 the National Front, despite standing for 300 seats, won only 1.3 per cent of the votes. The prime minister, Margaret Thatcher, had stolen their clothes using the language of being 'swamped' in relation to immigrants. But the overwhelming reason that the tide was turned against the National Front, both ideologically and on the streets, was the huge support from and diversity of ANL supporters. From this point on, anti-xenophobia and anti-racism became firmly rooted in the British labour movement.

2004 ONWARDS: WORKERS FROM THE 'NEW EUROPE'

Vacillating from Support to Scapegoating

From May 2004 up to one million workers from the newly joined EU countries of Central and Eastern Europe, two-thirds of them Polish, arrived seeking work in the UK in what was the largest single in-migration of people in history (Salt and Millar, 2006). Despite the EU's espoused 'fundamental freedom' of labour mobility, the UK was only one of three countries to fully open up its labour market to these new workers. Far from being motivated by altruism, the Labour government desperately needed to recruit workers to fill acute shortages of labour in what appeared to be a buoyant British economy. This new wave of migrants was different to previous ones; it was younger and more female, with the vast majority of arrivals aged between 18 and 34 and nearly half of them women.

In 2004 this inward migration was enthusiastically championed by the then Labour government. A Home Office (2007) report, citing the support of the Institute of Directors and British Chambers of Commerce, rejected any link between immigration and depressed wages or increased unemployment. This was bolstered by speeches from ministers in the Labour government. The home secretary, Jacqui Smith (2007), spoke of the 'purity of the macroeconomic case for migration', and the minister for immigration, Liam Byrne, argued that 'there are obviously enormous benefits of immigration ... There is a big positive impact on the economy which is worth £6 billion' (House of Lords, 2008: 22).

However, this positive mood music on migration came to an abrupt halt after 2008 when the global economy plunged into crisis, and it was clear that the apparently booming British economy was built on a mountain of bad debts. Particularly after 2012, as austerity started to bite, with a plunge in real wages and savage cuts to public spending, the narrative shifted from one of the positive contribution of migrants to them being viewed as a drain on public resources. Migrant workers became a convenient scapegoat for falling living standards and the anti-immigration propaganda of a reinvigorated United Kingdom Independence Party (UKIP) found an audience in some sections of the working class. This was reflected in UKIP breaking through in electoral politics in the 2013 local elections, the 2014 European elections and the 2015 general election.

Far from challenging ideas that blamed migrant workers for a range of economic ills, the Labour Party leadership (who were out of office) acquiesced to this populist rhetoric despite extensive research demonstrating, clearly and consistently, that recently arrived migrant workers contributed more in tax than they cost in public spending as well as filling shortages at the bottom of the labour market (see Ruhs and Vargas-Silva, 2018). The Migration Advisory Committee (2018) reviewed twelve studies conducted between 2003 and 2018 and found that immigration had little or no impact on average employment or unemployment of existing workers. In fact, in the immediate aftermath of the arrival of migrants from Central and Eastern Europe, from 2004 to 2008, real wages were rising; this came to an abrupt halt after 2008. It was the crisis and austerity that caused wages to fall, not the arrival of new workers. Despite an abundance of evidence in the public arena pointing to the dependence of the British economy on migrant workers and their positive contribution, the Labour Party in opposition chose to use apologetic discourses that played directly into the hands of xenophobes.

(Mostly) Positive Noises from the Top of Trade Unions

The responses of trade unions in more affluent EU countries likely to receive migrant workers were at best cautious and at worst hostile to the free movement of labour, although there were variations within and between unions. The German DGB (German Trade Union Federation) and its Austrian counterpart the ÖGB[2] opposed opening up labour markets, claiming not to be anti-immigrant but using the excuse that an influx of cheap labour would put pressure on collective agreements and labour standards. In contrast, the British TUC and its affiliated unions were more positive.[3] In the period just before 2004 the TUC produced leaflets in ten languages explaining the role of trade unions and how to join that were sent to workers who registered under the Worker Registration Scheme. This response was the cumulative effect of policies fought for in the labour movement, the history of self-organisation of black workers and the experience of a small section of the TUC that had worked with newly arrived Portuguese migrant workers in 1986. Proactive policies within large general unions such as the GMB and Unite were driven by socialists and anti-racist activists from below, but from a leadership perspective this stream of new workers on the labour market, mostly working in areas that they represented, offered the chance to boost their recruitment.

Some employers were quick to take advantage of inadequate English language skills and scant knowledge of employment entitlements to impose harsh working conditions and pay wages less than the legal minimum. The reality for many workers from Central and Eastern Europe, even though working legally, was one of exploitation and abuse. Complaints included excessive working hours with inadequate breaks and no enhanced overtime. Recruitment and temporary labour agencies were exposed for imposing exorbitant charges for finding employment, lower payment than promised and withholding wages. Tied accommodation, where bosses provide housing, resurfaced and many complaints from migrants were about overpriced, overcrowded and shoddy accommodation.

The recruitment and organisation of newly arrived workers from Central and Eastern Europe presented hurdles for trade unions. Large numbers of migrants were concentrated in the private sector and in agency employment where union membership was either low or non-existent. As with previous waves of migrants, language barriers, aggressive and avaricious employers and stretched union finances were challenges. However, though uneven there were many positive noises from the upper echelons of trade unions voicing support for working with new migrants. The small Bakers, Food and Allied Workers Union (BFAWU) actively worked with recruiting and supporting migrant workers and one of their executive officers pointed to continuities with previous migrations: 'My view on this new migration is what is new? When people came over here from India and Pakistan, we dealt with it ... So why can't we do it with the Poles. It's just a question of having in place structures and policies which enable you to do that (BFAWU executive officer)' (Fitzgerald and Hardy, 2010: 136).

The majority of trade union officers suggested that compared with the 1960s and 1970s unions were better at promoting solidarity between workers and not just 'workers who look and sound like us'. An executive officer of Unite said:

> It's partly our determination to be a champion of migrant workers and it's also our determination to say there are no no-go areas for trade unions in the twenty first century. I think it is essential that as a trade union we not only make sure that all are treated equally, but that we also speak out in favour of migration, making sure that others like Migration Watch do not fill the space left. (Ibid.: 137)

In 2005 the TUC set up a development fund for projects with newly arrived migrant workers. In 2006 a national document, *Organising Migrant Workers: A National Strategy*, was launched to concentrate the TUC's efforts to organise, represent and support migrant workers. Another dimension of trying to engage with these newly arrived workers was international collaboration, particularly between Polish and British unions, that involved the TUC signing a protocol with the two main Polish unions, Solidarity and OPZZ – quite a feat given historic vitriol between these two unions.[4] While the agreement was well intentioned, it had very little impact on the ground; the young age profile of the Poles in Britain meant that they did not have the historic memory of older workers. A much more concrete initiative was the secondment of Tomasz – a Solidarity national organiser from Gdańsk in Poland – who worked directly with trade unions, mainly in construction, in the North East of Britain to recruit Polish workers. Another success was the deal that BFAWU signed with an agency that recruited Polish workers giving them negotiating rights and that included a broader remit of auditing the accommodation provided by the agency.

Organising against Xenophobia on the Ground

Supportive policies and good resolutions from the top of the trade union movement opened up a positive ideological space and resources for working with migrant workers from below and it was the activity at the grassroots level that delivered imaginative initiatives. This included trade unions networking with a wide range of community organisations such as Polish clubs, law centres, Citizens Advice Bureaux and local churches to tackle abuses at work. In East Anglia the GMB union organised a fishing trip for Polish and British workers to combat antipathy towards Polish workers, who ate the fish they caught rather than throwing them back in line with the protocol of British coarse fishing.

In Southampton the GMB established a Polish-speaking holding branch set up in response to the demands of the Polish community. The organiser told me that they were expecting a handful of people, but at the first meeting in August 2006 over a hundred people crammed into a small pub room. By 2008 it had grown from 50 to 500 members and produced a layer of Polish activists and full-time organisers. This acted as a catalyst for the recruitment and organising of other workplaces. In one case management at a workplace that employed a high number of workers from the new EU countries would not listen to grievances on health and safety, so 20 workers joined the union and 55 put their name to the grievances. The GMB organiser told me that, 'We went in [to the other factory] and recruited 40 people in one day – Latvians, Lithuanians, Russians as well as Poles. The key to recruiting these workers was a Latvian woman. Had she not been there we would not have been so successful.'

There are not many precedents for this, but it is worth noting that at the beginning of the twentieth century the Socialist Party of America created seven foreign-language federations that successfully mobilised recent migrants.[5] Separate trade union branches are not a long-term solution but may be a mechanism for organising workers and workplace activity as a step towards unity between British and Polish workers.

However, it was not just a question of recruiting migrant workers but of overcoming obstacles to retaining members and encouraging them into activity. One trade union organiser told me that it was not difficult to run a campaign that showed that 2,000 people had joined, what was difficult was building a sustainable membership. This was illustrated when I interviewed four Polish shop stewards from a banana-packing factory in Luton where the poor working conditions had made national newspaper headlines. Their enthusiasm and support from an articulate and feisty union organiser from the Czech Republic was not enough. The management took advantage of the lack of language skills of shop stewards to consistently undermine them by refusing to hold meetings and ignoring grievances. The result was that a lack of tangible improvements by the union meant that it was difficult to retain members and to maintain an active workplace branch.

It would be wrong to assume that workers from Central and Eastern Europe are siloed in non-unionised parts of the economy. Where they were in workplaces with established trade unions, they have been on strike alongside British workers. In December 2005 a strike took place at the Iceland distribution depot in Enfield, North London, over pay and management bullying. Some of the placards on the picket line read '*Strajk Oficjalny*' ('Official Strike'), reflecting the large number of Polish workers involved. In 2007, when Polish agency workers were bussed in to try and break a Post Office strike over privatisation and pay in Watford, pickets climbed on to the bus and explained the dispute, and the Polish workers voted not to scab. Some young workers, disillusioned with the lack of support from mainstream unions, organised themselves. In 2018 an English–Polish-language agitational paper (see *Workers Wild West*, 2018, for example), mainly aimed at migrant workers in West London (Park Royal, Greenford, Heathrow), documented their lived experiences of working in the warehouses of giant supermarkets, in food-preparation factories and for agencies. It looks at how these workers have tried to organise themselves and their acts of rebellion in the workplace to get better pay and conditions.

Solidarity Is Not Automatic

Unite, the second largest trade union in Britain, illustrates the contested and contradictory nature of trade unions with regard to migrant workers. From 2004 onwards they organised many initiatives, funded by the union itself or the Union Learning Fund,[6] to engage with migrant workers by arranging popular English for speakers of other languages (ESOL) classes (see Chapter 9 on Sports Direct). But in the same union, in January 2009, 6,500 construction workers at the Lindsey Oil refinery came out on unofficial strike under the slogan of 'British Jobs for British Workers' (borrowed from the then prime minister, Gordon Brown). They were protesting against a subcontractor employing 'posted' workers from Italy, Poland, Portugal and Spain, on pay rates that were lower than the sector's national agreement between employers and unions.

The innocuous-sounding Posted Worker Directive, introduced in 1996 by the EU, is a pernicious recipe for classic divide and rule. It allows employers to temporarily 'post' workers in another country, and although employers are obliged to pay the minimum laid down by the host country this is often well below the rate agreed nationally between employers and trade unions. The metaphor of divide and rule became literal when the Italian construction contracting company, IREM, in the Lindsey dispute created a barrier between migrant and indigenous workers by accommodating their workers in large, grey housing barges moored off Grimsby docks. It is not the case that the dispute was overtly xenophobic. When other Unite activists and socialists visited the picket lines the slogan was quickly replaced with one that read 'Fair Access for Local Labour' and the British National Party (fascist) and UKIP (right-wing populist), who sought to capitalise on the dispute, were thrown off picket lines. However, this 'moment' demonstrated the potential for division between indigenous and migrant workers.

HERE TO STAY, HERE TO FIGHT

As Yuri Prasad (2017) points out, whether the hold of racism or xenophobia on white workers can be broken is not predetermined or a matter of chance. As we have seen, it depends on the commitment and actions of a minority of workers within the working class who campaign for policies against discrimination in their unions and challenge xenophobia in the workplace. The story of the transformation of the working class is one that has been repeated throughout the history of capitalism in which the movement of workers within and between countries is a constant feature.

The shift of trade union bureaucracies in Britain to embrace diversity and condemn racism is to be welcomed and constitutes a sharp contrast to the poisonous environment of the 1950s and 1960s. For example, the TUC statement encouraging members and individual unions to join the Stand Up to Racism march in November 2018 is a measure of how far the official labour movement has come in six decades and is something to be proud of.

Most British trade unions have excellent policies against discrimination and for promoting diversity as well as sections for black members. But these do not automatically translate into the elimination of xenophobia and racism in the workplace that is so deeply rooted in the history of British capitalism. Scapegoating migrant workers in times of crisis is not just the preserve of populist and right-wing parties. As we have seen, the Labour Party and some trade union leaders have made concessions to this populist rhetoric in a more coded way rather than challenging it. Socialist and anti-racist activities have been vital in fighting against divide and rule in trade unions and the workplace, but, as we shall see in Chapter 9, strikes and campaigns of cleaners are inspiring examples of the self-organisation of migrant workers taking action to win stunning victories.

9
Taking the Bosses to the Cleaners

DAVID AND GOLIATH STRUGGLES

The Dirty Work of Neoliberalism

Cleaners are an invisible army of workers labouring at the crack of dawn and late in the evening. Night buses are not just to get revellers home in the early hours, but to transport cleaners through the arteries of big cities. They clean the detritus of government departments, the richest global corporations and universities where vice chancellors are paid £300,000 and above. The contrast between these badly paid workers and the bloated salaries of those whose offices they clean is a stark reflection of inequality in Britain. However, these are not workers in some criminal underbelly of British capitalism, but the employees of lucrative global service sector multinationals. The industry is characterised by a bewildering merry-go-round of firms, with mergers, takeovers, changes of names and contracts being ditched and passed on with the result that cleaners are bounced from one contract and employer to another like so much human cargo. By subcontracting badly paid work, government and educational organisations keep their hands clean by delegating the details of exploitation – passing on their dirty work both metaphorically and literally.

A stone's throw from the Houses of Parliament is a massive 1960s office block that is home to the British government's Department for Business, Energy and Industrial Strategy (BEIS). At midday on 15 July 2019 the cleaners and caterers were clapped out of the building as they walked out on indefinite strike. One striker, Gloria, told me they had no idea how long they would be out for, but had no expectations that it would be quick and were 'prepared to stay out for as long as it took'. Among the strikers were workers who had just started and others who had worked there for 24 years. This chapter looks at the struggles of cleaners in the new millennium at BEIS and the University of London. With low pay, often no savings and living from one pay cheque to the next, their financial fragility made taking industrial action a tough decision. Challenges in trade union organisation in the cleaning sector were compounded by the fact that migrant workers dominate the case study workplaces; in the University of London cleaners are mainly from Latin America, while in the government department, BEIS, they come from Europe, the West Indies and Africa. These workers were on a patchwork of different contracts: while the working conditions of existing employees were sometimes protected, newly recruited workers were nearly always employed on worse conditions. In David and Goliath disputes, despite these organisational barriers and ruthless tactics from employers – and sometimes disinterest from established trade unions – these cleaners have taken on giant service sector multinationals and powerful institutions in higher education and government and won stunning victories. This section begins by looking at landmark struggles by cleaners from the 1970s before looking at more recent struggles.

Struggling against Subcontracting: 1970 to Early 2000s

In the rush to attribute all changes in employment to the neoliberal era, cleaners have been bracketed into the binary divide of direct, regulated work and subcontracted precarious contracts. Yet the picture in the 1950s and 1960s was far more uneven. With the growth of large offices in the 1930s, the first modern cleaning companies were established and a significant tranche of cleaning was contracted out to these firms. The then Labour government started subcontracting in the public sector as early as 1968 when it made cuts in the civil service, sacking 4,000 directly employed cleaners. In contrast to the image of work in the 'golden era' as one of stability and a job for life, May Hobbs, the energetic leader of the campaign to unionise cleaners in the early 1970s, paints a very different picture of employment in the East End of London (Hobbs, 1973). She describes pawn shops as a constant feature of day-to-day existence and families living on the edge of plunging into debt and eviction. Before becoming a cleaner, May Hobbs had a succession of different jobs, working variously in shoe and sweet factories and then in a workplace that made cardboard boxes.

In the 1970s, from a trade union perspective, these women workers were invisible. The TGWU (or T&G, now Unite),[1] born from the unionisation of unskilled workers, was largely white and male and cleaners hardly featured on their radar. Sheila Rowbotham describes the role of the women's liberation movement in supporting the cleaners: 'In the autumn of 1970, a crowd of women and one man packed into my bedroom in Hackney, East London, to hear May Hobbs tell us about her efforts to organise cleaners. The Night Cleaners' Campaign had begun' (2006: 609).

Unable to track down anyone from the T&G and even less able to find their branch meetings (to the extent that May Hobbs took to phoning the general secretary, Jack Jones, and complaining to his wife), the campaign approached the Civil Service Union (CSU). They already had some cleaners as members and were described by Sheila Rowbotham as having 'a markedly different style of trade unionism' (ibid.: 613) with a 'young and zippy' union official. In the summer of 1972, cleaners in two Ministry of Defence Buildings – the Empress State and the Old Admiralty – went on strike. They got instant support from other trade unions: T&G members would not cross picket lines to deliver supplies and Post Office workers refused to deliver mail; printers, railway workers and clothing workers sent donations. Targeting high-profile government buildings brought results. The CSU was able to get the contractors to recognise the union and the strikers won a pay increase of £2.50 a week and a 50 pence night allowance. The CSU then began to press for the cleaning of government buildings to be taken back in-house. Although this was a significant victory the campaign did not maintain its momentum. It did, however, inspire other attempts to organise cleaners.

The election of the Conservative government in 1979, with Margaret Thatcher as prime minister, represented an ideological shift. Proselytising the benefits of the market, privatisation became a government crusade and there was a step change in contracting out services in the public sector. This enabled big players in the cleaning industry to transmogrify into service sector multinationals, bringing cleaners to the forefront of industrial struggle. Cleaners in the public sector did not necessarily earn higher wages, but crucially they had the safety net of receiving sickness benefits, holidays and pensions. In the summer of 1995 low-paid ancillary workers, mainly East Asian women, from Hillingdon Hospital in West London went on strike after their contract cleaning employers, Pall Mall Health Support Services, cut their already low wages by 20 per cent, from £3.20 to £2.50 an hour. After being outsourced in 1986 (one of the first private contracts that was forced through under Thatcher's Conservative government) they had already lost sick pay, bonus and pension rights and the workforce was cut by a third. In an attempt to intimidate the women in advance of issuing new contracts, the firm demanded the women's passports and questioned their immigration status (Ransom, 1997).

Fifty-three women refused to sign the new contracts and were locked out; the strike began on 1 October 1995. Initially supported by Unison (a giant union representing public sector workers), negotiations between officials and management produced an offer of £350 to compensate the women for their jobs. But for those on strike this missed the point; what they wanted was their jobs back on the same terms and conditions. Unison withdrew official support and from January 1997 stopped their strike pay, but nevertheless 31 women decided to stay out and fight on. In another roll of the dice the contract was passed on yet again – this time to Granada, another giant in catering and media. The women eventually won two industrial tribunals against Pall Mall and its successor, Granada, which ruled that they must be reinstated with maximum compensation. In this long dispute, despite not receiving official support from the trade union movement, the women received solidarity from workers all over Britain and internationally. They spoke at meetings and assemblies in many countries and developed links with other groups of workers, including marching alongside the Liverpool dockers and their families striking against casualisation in 1995. One activist related how, as the National Union of Journalists' mother of the chapel[2] in a Midlands university, she organised a levy of members and invited the women to speak; branch members attended the joyous final curry and Punjabi drummer bash of the strikers in Hillingdon.

As was outlined in Chapter 6, drawing on campaigns in the US, in the new millennium the issue of poor pay for cleaners was raised when a community initiative, TELCO, launched a campaign and won the living wage for groups of cleaners. However, unlike the broad alliance of trade unions, faith and community groups, and enlightened employers that characterised TELCO, as we shall see in the next two sections, trade unions and industrial action were at the centre of the struggles of cleaners and ancillary workers at the government BEIS department and the University of London.

HYPOCRISY AT THE CENTRE OF GOVERNMENT: ALL-OUT STRIKE AT BEIS

Prominently displayed above the reception desk of the government's BEIS department are slogans proclaiming its aims as: 'Creating a world class business environment' and 'Improving pay and working conditions nationwide'. These grandiose claims could not be further from the treatment of their low-paid cleaners, post and security staff, porters and receptionists who were outsourced to Engie, and those working in hospitality who were outsourced to Aramark. These were both global transnational firms that paid minimum wages and imposed poor working conditions. Even more risible is that fact that BEIS commissioned, published and promoted the Taylor Report that investigated poor working conditions in Britain. One clause in its terms of reference reads:

> Because, despite the important contribution of the living wage and the benefit system, fairness demands that we ensure people, particularly those on lower incomes, have routes to *progress in work*, have the opportunity to *boost their earning power*, and are treated with *respect* and *decency* at work. (Taylor Review of Modern Working Practices, 2017: 6, my emphasis)

Yet the lived experience of these outsourced workers at BEIS showed that 'decency' and 'respect' were in short supply. The PCS branch secretary described the long hours worked: 'Security guards are contracted to do a 60-hour week – that is five twelve-hour shifts. But cleaners regularly work two shifts in one day just to make ends meet. By working from 5.30 to 8.30 and then 10.30 to 6.30 they do an eleven-hour day.'

Gloria, one of the caterers, described how time away from home was stretched even further because she couldn't afford to travel on the London Underground during peak times: 'I have to get up an hour earlier to get the bus than I would if I used the tube – but to save money we can juggle the bus pass in the family if we are doing different shift patterns.'

The Campaign and the Trigger

The campaign for better terms and conditions started in May 2017 when the PCS branch committee decided to try to recruit outsourced workers. One branch officer explained how they decided to 'mass recruit and do something. Let's move so that people can see we are serious ... why should they give us £7 a month [subscription] that they can't afford in return for a vague promise?'

But despite trying to tap into this pool of mainly non-unionised outsourced workers, recruitment was negligible. It was in September 2018 that the campaign began in earnest with a small and lively lunchtime demonstration with PCS activists teaming up with the relatively newly established UVW who were organising outsourced workers at the nearby Ministry of Justice. The protest took place without seeking permission or formal approval from union officialdom or the police. The demands of the PCS campaign were for the London living wage, parity of conditions with directly employed workers and for jobs to be brought in-house. An online petition gathered thousands of signatures, template letters were produced for members to submit to put pressure on the employers and the branch committee drafted a detailed report demonstrating that bringing outsourced jobs back in-house was more cost effective. Staff events at BEIS were door-stepped by branch activists with leaflets to raise the profile of the campaign.

In autumn 2018 and the beginning of 2019 the poisoned chalice of an increasingly militant and determined group of workers was passed between the subcontractors and BEIS management. The branch secretary argued that BEIS had colluded in perpetuating low pay: 'It was an eye-opener when we found out that the subcontractors – Engie and Aramark – had been told by BEIS to align the pay of their staff to the average for the private sector in London – a rate far below the London living wage.'

Meanwhile, the secretary of state tried to kick the issue into the long grass by announcing a review of terms and conditions. However, in the face of stalling by BEIS management and the subcontractors, the union achieved a 100 per cent 'yes' vote in a ballot and a one-day strike was held in January 2019. This was followed by a two-day strike in February alongside smaller unions, the IWGB and UVW, from other government departments. When workers saw that the union was serious about trying to improve working conditions and pay there was a step change in recruitment to the PCS at BEIS.

In March 2019 the contract for cleaning was bounced from Engie to ISS (a Danish company owned by Lego) which described itself as: 'a leading workplace experience and facility management company. In partnership with customers, ISS drives the *engagement* and *well-being* of people, minimises the impact on the environment, and protects and maintains property' (ISS, 2019, my emphasis)

This lofty statement was in stark contrast to the contempt with which ISS treated its workers when, within a few weeks, they unilaterally decided to 'upgrade' and 'harmonise' its payroll in what was labelled 'Project Greenfield'. Workers who had worked for three weeks would be paid for just two on the new payday and would not receive the money that they were owed until they left the company. Withholding one week's wages from workers who lived from one pay cheque to the next imposed real hardship and distress on ISS workers in BEIS and all over Britain. This was far from the experience of the top management of ISS whose five UK directors shared remuneration of £2 million for the year ending December 2017, with the highest-paid director earning more than £500,000 (Quinn, 2019a).

Tensions were ratcheted up even more in BEIS when pay packets were repeatedly miscalculated and nearly always underpaid. Management had to give some members of the branch committee 100 per cent facility time to rewrite the pay slips of the workers employed by ISS to ensure that their members were not short-changed and put under financial duress. Workers desperate to contact the firm found that ISS's much-vaunted helpline and email query system went unanswered. One worker, Carlos, explained that: 'what really triggered me was there was no contact, no communication [from ISS] ... I felt cheated, not respected as if I didn't matter ... it would have been a minor dispute if they had apologised'. This was the match that ignited the tinder of accumulated grievances; after May 2019, PCS union membership among outsourced workers leapt up.

Onto a War Footing and Winning the Battle

A food bank was set up in front of the office of the secretary of state and on each of the ten floors of the building. This had the effect of building solidarity with office workers who overwhelmingly supported the action, embarrassing top civil servants and the secretary of state and providing some workers with a necessary lifeline. But the key to pressuring the bosses was industrial action. The PCS branch reballoted its members in March 2019 and in May 2019 cleaners and caterers – the strongest groups – walked out on strike for one week. In June all outsourced workers came out (see Figure 9.1). Strike pay of £40 a day was provided by the PCS nationally and the local branch topped this up with large donations from office workers in the branch working in the same building. Refusing to be fobbed off any longer, the branch escalated the action and on 15 July 2019 the cleaners and catering staff walked out on indefinite strike. One committee member admitted to some trepidation: 'this was a helluva thing to do … a few months before none of these workers had been in unions and most of the cleaners didn't speak English!'

Figure 9.1 Demonstration of striking members of the PCS union at the Department of Business, Energy and Industrial Strategy in June 2019
Source: Guy Smallman.

No cleaners from BEIS were going into work and although agency cleaners were redirected from other buildings, rubbish and dirt built up. Catering was completely shut down and a branch committee member explained: 'What really p****ed off management was shutting down catering. They make lots of money from letting space out for conferences and with no catering people were cancelling all over the place.'

In September the receptionists asked to be balloted and went on strike (one week on and one week off). Senior civil servants drafted in 'to work reception' ended up with a queue of people stretching out of the building and down the road. Solidarity from other unions and groups of activists was really important for the strike. The RMT donated £5,000, and there were visits from left-wing Labour MPs such as John McDonnell and Rebecca Long-Bailey and even the TUC general secretary, Frances O'Grady. The picket line had a festival atmosphere with striker Victoria enthusiastically describing that, 'There were noisy picket lines everyday with music and dancing … the cook-out of "jerk chicken" was really popular … there was lots of support from other unions, especially the FBU who allowed a rally to be held on top of a fire truck.'

Workers were transformed by their experience in these disputes. Strikers from BEIS talked about how, travelling round the country and speaking at meetings for the first time in their lives, their confidence had grown. Winnie, who worked in catering, described how exhilarating it had been to be one of a group of strikers that travelled to (Royal) Tunbridge Wells (a Conservative heartland in Kent), the constituency of the secretary of state for BEIS, where they held a noisy street meeting. This story of pride and increased confidence came through from all the workers interviewed.

The strikers refused to back down. The bosses never imagined that they could hold out for so long. Twelve weeks after the start of the strike, by the end of October, BEIS, ISS and Aramark agreed a deal with the PCS. They won a massive victory, achieving the 'gold standard' in pay and terms and conditions. The deal they won included receiving the London living wage as a minimum, annual pay rises, occupational sick pay on full pay and up to six extra days' annual leave as well as increases in the overtime rate. The actions of only 25 cleaners and 17 catering staff won these hugely improved conditions for all outsourced staff in the building, a deal that was then rolled out to ten smaller BEIS departments and agency sites across the country. The victory not only transformed the confidence and pride of the strikers but made a material difference. Cynthia described how the extra money in her pay packet had changed the quality of her life:

> Before I was living hand to mouth ... some months I would have to skip a bill. But the extra goes a long way ... My mum in Jamaica was ill – you have to pay for medical care there – I could send some money home ... I can put some money in my pocket and go out with friends – even treat them. This year was the best Xmas me and my kids had ever had ... for once I didn't have to skimp.

THE DIRTY WORK OF LONDON UNIVERSITIES

The Bloomsbury campus of the University of London has been the site of a number of battles at the School of Oriental and African Studies (SOAS) and the Senate House administrative building where outsourced workers have demanded and won a living wage, better terms and conditions and that services are taken back in-house.

SOAS

The dispute of outsourced cleaners at SOAS was inspired by success at Queen Mary, University of London in winning the minimum wage and getting brought back in-house in 2006. Getting organised was provoked not only by the minimum wages and poor terms and conditions of workers at SOAS, but also by the fact these highly profitable multinational subcontractors were often in arrears paying wages. The campaign by the SOAS Unison branch for the London living wage was launched with a screening of Ken Loach's film *Bread and Roses* (2000) documenting the successful unionisation of Latin American cleaners in Los Angeles. The Unison branch secretary at SOAS, Sandy Nicholl, describes how the campaign was built through 2007 and 2008: 'We held public meetings, had an online petition, a teach-in, a May Day march through Bloomsbury and a lobby of the governing body. In June 2008 we won the London living wage and the pay of the cleaners increased by 34 per cent – but bringing the cleaners back in-house was a battle that we had yet to win.'

The campaign continued and the cleaners won union recognition with the contractor ISS the following year. But the employers had their revenge for being forced to make this concession. Sandy goes on to describe how:

> Only three days later, on 12 June 2009, ten cleaners were taken into detention and six of them deported within 24 hours, after a dawn raid by the immigration officers called in by cleaning contractor ISS – even though these cleaners had been employed for many years. Cleaners were told to attend an 'emergency staff meeting' at 6.30 am – this was used as a false pretext to lure them into a closed space from which the immigration officers were hiding to arrest them. More than twenty officers, dressed in full riot gear, aggressively interrogated the cleaners and then escorted them to the detention centre.

Further intimidation followed when the Unison branch chair, an Ecuadorian law graduate, Stalin, who worked in the post room and was central to the campaign, was sacked on trumped-up charges of misconduct. The determination of the employers to break the union organisation of cleaners was clear when his arbitrary dismissal was upheld despite official backing from Unison, a 24-hour strike that closed the campus and dozens of individual character references. Sandy Nicoll was clear that these actions were designed to:

> Break strong and powerful union organisation on the campus and put the cleaners back in their place. The SOAS management were complicit. But this did not have the desired effect of intimidating the workers. Union recruitment increased and spread to catering and security and by 2013 there was 100 per cent membership with seven trained representatives in Unison.

In 2014 the campaign for bringing these jobs in-house was ratcheted up with a 48-hour strike in January that shut the whole campus (see Figure 9.2). Trade union delegations came from all over London and other union branches up and down the country sent workplace col-

Figure 9.2 Cleaners at SOAS begin a 48-hour strike in March 2014 after a 100 percent vote for action in the fight over sick pay, holidays and pensions
Source: Guy Smallman.

lections. Workers achieved an important win when sick pay and other conditions were harmonised with directly employed workers, including an extra ten days' paid holiday. But the momentous, final victory came in August 2018 when, after an eleven-year battle, employers agreed to bring cleaning and other services back in-house. Outsourced workers not only benefitted from full university pay and conditions, but also opportunities for training and development that had previously been denied to them.

Senate House

Inspired by the SOAS workers' success in winning the London living wage in 2008, in July 2011 outsourced workers at Senate House (the administrative centre of University of London), a one-minute walk from SOAS, started a campaign for the London living wage. In the 1980s there was a history of Latin American activists who brought radical traditions with them and successfully unionised a West End store. They went on to be involved in the TGWU (T&G) and developed their own section in the union with a newsletter called

El Mopo (The Mop) (Hearn and Burgos, 2010). However, Henry, a leading organiser of the workers at Senate House, did not have a history of political involvement – he claims that he 'didn't know anything about trade unions'. The inspiration was reading about other workers who had won the London living wage, and what galvanised him into organising his co-workers was the poor treatment and low pay of cleaners in Senate House. With a handful of other trade union activists he set about the daunting task of visiting cleaners in all the halls of residence and systematically recruiting them to the union. Henry explained that, 'We organised ourselves into the Unison Senate House branch and quickly won some important victories ... in 2011 we got payment of overdue wages sorted and then we got union recognition from the [then] cleaning contractor Balfour Beatty ... in 2012 they agreed to pay the London living wage.'

The recognition agreement meant the right to collective bargaining and time off (facility time) for elected representatives. As a result, in a short time the branch grew by two-thirds and recruited outsourced workers, many of whom were migrants with a Latin American background. This was an achievement given the organisational complexity of outsourced workers who were employed by four different multinational subcontractors on a multitude of contracts. In September 2012 the branch launched a new unifying campaign that demanded parity with workers who were directly employed in terms of sick pay, holiday pay and pensions. But this time the campaign did not receive support from Unison and by February 2013 serious tensions had emerged between these workers and Unison leaders. Henry (then Unison representative and now president of the IWGB) explained that: 'We couldn't wait for the bureaucracy ... we had problems with them ... let's take over the branch to reflect the membership and we put up a cleaner as vice-chair.'

Unison invalidated the result of the elections on what Henry argues were spurious grounds. The migrant cleaners frequently changed their accommodation and many ballot papers had to be reissued; this was the pretext used by Unison. The cleaners held a demonstration outside Unison headquarters and the police were called. In protest, in 2012 70 of the cleaners, caterers, security guards and other maintenance workers left the Unison branch and joined a newly formed rank-and-file union on campus – the IWGB. The intransigence of the employers in recognising the IWGB did nothing to diminish an enthusiastic and militant campaign that mobilised external support on the campus from students and academics and drew on organisations from the Latinx community in London. A two-day strike in November 2013 won up to six months' sick pay, three days' holiday and access to the pension scheme. Spurred on by these gains strikes took place in January 2014 to fight for pension rights and union recognition: the 'Bus for Justice' involved a high-profile parade through the heart of London.

But these lively and direct tactics led to friction with the employer, the University of London, claimed that it would not negotiate with a union using 'direct action and adversarial tactics'. The IWGB accused Unison of working with the employers and, according to Alberti (2016), Senate House files seized by students in an incursion into the vice chancellor's office suggested a tightening partnership between the university and the recognised trade union, Unison. There have been 17 days of strike action since September 2017, when University of London cleaners and security began their campaign to end outsourcing, with calls for a boycott of Senate House beginning in December 2018. Impressively, the IWGB, whose membership and resources are a fraction of those enjoyed by Unison, managed to give their members full strike pay through funds raised with parties, crowdfunding and raffles.

SELF-EMANCIPATION AND POLITICS: THE INGREDIENTS OF SUCCESS

These campaigns and strikes raise important questions for activists about the organisation and self-organisation of low-paid migrant workers. Cleaners at the University of London and government department BEIS won stunning victories against outsourcing as well as substantial improvements in pay and conditions. This was despite employers tapping into what they wrongly assumed was the passivity of migrant workers, trying to exploit both their lack of English language skills and knowledge of employment rights, or, in the case of SOAS, vindictively wielding the stick of the UK Border Agency.

Workers were organised and self-organised and took action in a variety of different trade union formats, including mainstream unions, both large (Unison with 1.3 million members) and medium-sized (the

PCS with 200,000 members) and new unions such as the IWGB. A decisive element in each case was the role of branches and the involvement of members and, in particular, the role of activists within those branches, often branch officers. In BEIS and SOAS the campaign was initiated by the branch, which then kept control of the pace of the dispute. In particular, the energy, commitment and accountability of lynchpin branch secretaries – both socialists – were central to sustaining and winning the disputes. At BEIS the PCS at a national level stepped in and supported the action, critically by paying full strike pay and ensuring the presence of full-time union workers on the picket line. The strikers spoke highly of the support that they had received from PCS. Pierre said, 'They listened to us ... the support was awesome ... they made sure we got a deal – we couldn't have got better.' Other workers described the branch secretary as 'brilliant' and praised the way that there was complete transparency at every stage.

Building the PCS union in BEIS and Unison in SOAS was a dynamic process, not a mechanical, linear progression of recruitment then taking action when there were 'enough' members. Rather it was the reverse – it was activity and the promise of doing something that persuaded workers to join and built the union branch. In the case of both Unison and the PCS this led to new activists being trained as union representatives/stewards: in the Senate House dispute the leading activist, who had left Unison and taken members with him, went on to form the IWGB.

In SOAS politics was key. The presence of union activists who were socialists and the wave of student activism from 2010 that engaged them with workers' struggles were pivotal to the militancy, vibrancy and tenacity of the campaign. The campaign and action were sustained over nearly a decade, during which time the branch was unremitting in putting pressure on the employers. Sandy Nicoll explained how building coalitions with students (National Union of Students) and the lecturers union UCU was central to the success of the campaign: 'It was really important having the Justice for Cleaners campaign to support the action of the cleaners. The students were respectful of the fact that the cleaners were in charge – it was their dispute – but they could do actions like occupations that we couldn't do as a [Unison] branch.' He went on to argue that the role of politics was writ large in the success of this campaign and overcoming the limits of the trade union bureaucracy:

> It was not a narrow trade union dispute – it was interlinked with questions of migration and racism that had to be taken up. Unity between the students and cleaners was key – 30 cleaners came on the student demonstration against the hike in fees in 2010. Socialists in the branch supported the outsourced workers and played a role in shaping the terrain of struggle, but at the core of the campaign and action was the self-emancipation of a group of workers.

It was the recruitment, mobilisation and involvement of workers from below and their zeal and willingness to take strike action that closed campuses and forced concessions from pressurised and embarrassed employers. In BEIS brave workers committed themselves to indefinite strike action that put the multinational employers and powerful institutions for whom they cleaned under duress. The momentous successes of the cleaners has given confidence to other groups of outsourced workers and Justice for Cleaners' campaigns. Between 2018 and 2020 Birkbeck College, Kings College, the London School of Economics and Goldsmith's have all brought their cleaners back in-house. Workers up and down the country, in hospitals, for example, have been inspired to fight against outsourcing in a rash of disputes and strikes that have barely registered outside the left-wing press. In Wigan, north-west England, in June 2018, after a series of strikes – a 48-hour strike followed by a five-day strike – hospital workers forced their NHS Trust to drop all plans for outsourcing cleaning, catering and domestic services that was a blow to the employer's plans for privatisation. These actions are a strong antidote to pessimistic assumptions that migrant and low-paid workers cannot win struggles. As we have seen, they have done so even when the odds appeared to be stacked against them. Far from being a 'blind spot' in the dark recesses of labour organisation these workers have turned out to be a beacon of hope for all workers.

10
Working and Organising in New 'Satanic Mills'

NEW SATANIC MILLS

Taking too long to go to the toilet, talking too much or taking time off to look after sick children would get you a warning. 'Six strikes' and you were out of a job. Workers queuing up to be searched after every shift, 'to prevent them from stealing goods from the factory', which kept them on the premises for up to an hour every day that they were not paid for. These are not stories of working in what William Blake's poem in 1808 described as the 'satanic mills' of the nineteenth century,[1] but accounts of workers' experiences in the Sports Direct warehouse in the twenty-first century relayed to me by a full-time Unite organiser. Sports Direct is the largest sporting retailer in Britain and its distribution centre is located on the 93-acre former site of the Shirebrook colliery (closed in 1993) on the border of Derbyshire and Nottinghamshire in the East Midlands. The warehouse employs some of the lowest-paid workers in Britain, mostly on contracts with no sick pay or pensions and subject to appalling conditions. The founder and chief executive officer of Sports Direct, Mike Ashley, has generated such a large stream of income from imposing notoriously bad work practices that he has been able to cannibalise ailing firms on the British high street.[2]

But Sports Direct is not simply a 'bad apple'. In July 2019, Amazon workers in Britain joined thousands of workers all over the world protesting against low pay and poor working conditions. A survey of Amazon warehouse workers in England by the workplace campaigning group Organise found that 74 per cent of people were afraid to go to the toilet for fear of missing productivity targets (Wolfe-Robinson, 2019a). Holding GMB union placards that read 'We Are Not Robots', demonstrations were held in the towns and cities of Rugeley, Swansea, Peterborough, Warrington and Coventry. In the warehousing sector competition spurs the constant drive to higher productivity, with low wages and the intensification of work. Despite the huge potential for organising in these new satanic mills that employ thousands of workers under appalling conditions, this does not automatically translate into trade union membership or taking industrial action. The case of the Sports Direct warehouse is a microcosm of the sector, illustrating how naked attempts to squeeze the pay and conditions of work plays out in one warehouse and providing an anatomy of workers' resistance and organisation.

WORKING IN A 'GULAG'

In 2015 I met up with a full-time Unite officer in a café in Nottingham, who described the battle against 'gulag' conditions in the Sports Direct warehouse. The culture of punishment and surveillance was reflected in a pernicious 'six strikes and you are out' policy. There was no procedure, no come-back and no defence against this draconian system. Locally, primary school teachers reported that children were sent to school when they were ill because their parents were too frightened to take time off work for fear of losing their job. Midwives in the area wrote a report voicing their concerns that pregnant women were not seeking advice early enough because of their reluctance to take time off work. 'Mr Loverman' – well known among the workers for his sexual advances and harassment of women – offered contracts or enhanced hours in return for sexual favours. Management had complete control to discipline or dismiss at will and any challenge to the system by workers was likely to result in the number of hours they were offered being reduced or withdrawn completely.

The compulsory searches carried out at the end of shifts meant that workers were paid below the minimum wage. There were punitive deductions from wages, such as docking 15 minutes' pay for clocking on one minute late on arrival or on returning from a break. The health and safety record of the Sports Direct warehouse was shocking. A Freedom of Information request by Unite to Bolsover District Council revealed that a total of 110 ambulances or paramedic cars were dispatched to the Shirebrook warehouse's postcode between January 2013 and April

2016. Fifty cases were classified as life-threatening, including chest pain, breathing problems, convulsions, fitting and strokes, and five calls were from women suffering pregnancy difficulties, including one woman who gave birth in the toilet of the warehouse. Yet no action was taken on the part of Bolsover Council or the Health and Safety Executive in response to this litany of incidents.[3]

This was far from the lofty promises that accompanied the arrival of Sports Direct, which was described as an exciting plan 'to breathe new life into Shirebrook'. At the launch the labour minister responsible for regeneration optimistically proclaimed that 'the jobs-boosting development would help local folks overcome the effects of the collapse of the mining industry'. The reality for the locality and the workers was very different. Sports Direct dominated the site by occupying half of the total space. Few local workers were employed, rather migrant workers, who were assumed to be cheaper and more compliant, were directly recruited from poor small towns and rural areas in Poland and later Romania. Sports Direct hoped that they could capitalise on the physical isolation of the workplace, and the lack of knowledge of legal entitlements and English of these newly arrived workers. The 'business model' used by Sports Direct built in complete flexibility for the firm at the expense of the workers. There were only 200 employees on permanent contracts with a reserve army of 4,000 other workers employed by two agencies – Transline and The Best Connection. Workers' accounts of these notorious work practices at the Sports Direct distribution centre, exposed by the media, prompted a House of Commons Select Committee (2016). Documenting a litany of abuses including an atmosphere of fear and bullying, pay below the minimum wage and an atrocious record of health and safety, the report makes shameful reading.

Subcontracting: From the 'Butty' System to Employment Agencies

It has been argued that the extensive presence of employment agencies that bundle workers out to different workplaces for varying lengths of time are a novel feature of the work environment. They wield disproportionate and arbitrary power in deciding what and how much work is available for individual workers, making it difficult for them to speak out. As we saw in Chapter 4, there has been a tendency to treat agencies as a completely new phenomenon, but subcontracting labour has been around for a long time. For example, the 'butty' system of supplying workers to employers has a long history in the area that can be traced back to the mining industry in the nineteenth century. Parallels with the modern-day subcontracting system through employment agencies are stark, characterised by fear, poor health and safety, nepotism and sexual harassment.

Workers signed up with one of the two employment agencies and then worked for the agencies' client, Sports Direct. The majority of workers had contracts from the agencies that guaranteed 366 hours a year. But these were front-loaded with workers typically employed on 40-hour-week contracts for nine weeks at the beginning of the year, after which there were no guaranteed weekly hours. In other words, they were left on ZHCs for the remainder of the year (House of Commons Select Committee, 2016: 60). The arrangement was completely one-sided and the House of Commons report quoted the agency Transline as insisting that they were not obliged to provide work. However, it was clear that workers were coerced into accepting any offer of employment: 'Refusal to accept a suitable assignment without a good case will result in you being deemed not available for work and may constitute gross misconduct. This may result in the termination of employment without notice and without payment in lieu of notice' (ibid.: 7).

The agencies found other ways to scam the workers and skim their wages. Transline offered pre-paid debit cards onto which their wages were paid. Workers were charged a monthly management fee for every cash withdrawal. Although the pre-paid system was administered by Cotis, Transline received £1.96 for each card user. The Best Connection agency charged a fee for 'insurance services' ranging for 45p to £2.45 per week that was deducted from wages. Workers were given no proper explanation as to what it covered (ibid.: 14). So poor were the procedures for recruitment and monitoring workers by the Transline agency that two Polish men were jailed for trafficking Polish workers to work in Sports Direct through their agency. These trafficked workers had their passports confiscated, were housed in horrendous and crowded accommodation and had most of their wages pocketed by their traffickers (Davies, 2017).

The warehouse is open 24 hours a day 365 days a year. The nature of the business can be forecast with some accuracy, so there was, and is, no economic reason for employing workers on short-term temporary contracts. Sports Direct and Mike Ashley have benefitted directly from subcontracting labour to agencies as it allowed them to drive down costs associated with the recruitment, payment of wages and deductions for tax and insurance – functions carried out by the agencies. Agency workers, some of whom had worked for the firm for more than a decade, were kept in a state of fear and insecurity. This arrangement was lucrative business for the two agencies which swallowed up around 10 per cent each of the total turnover of Sports Direct, adding up to £50 million a year between them (ibid.: 6).

THE CHALLENGE OF ORGANISING

I met Piotr in the Miners Social Club in Shirebrook. He was a stalwart of the Unite trade union and had been a representative for almost as long as the eight years he had been working at Sports Direct since he came over from Poland. Piotr described the uphill battles he faced with bosses in the warehouse as well as the vacillating support from Unite. From 2015 onwards, Unite mounted one of the most prominent examples of community unionism in a high-profile and lively campaign that engaged its own members and activists from the local community. This included trades councils in Nottingham, Mansfield and Chesterfield, branches of local unions (the NUT, the UCU and the PCS), Unite Community branches, Citizens UK and individual members of the Labour Party and Socialist Workers Party in the area.

The campaign had two prongs: first, to name and shame Sports Direct on a national level and embarrass them into improving conditions; and second, to build links with and organise the workers from the warehouse. The campaign was highly successful in exposing the appalling conditions in the warehouse, with strategic press releases and contact with the media leading to television programmes on the BBC and Channel 4. In September 2016 Unite Community branches organised protests, supported by the charity War on Want, outside Sports Direct shops in over 40 city centres. Unite established a relationship with the Newcastle United Football Supporters' Club, the team owned by Mike Ashley, and in January 2016 supporters of the club smuggled in and unfurled a banner saying 'SportsDirectShame' that was reported in the national press.

The second prong of the Unite campaign, pivotal to forging links with workers at the warehouse, was a two-year ESOL project set up by Unite in 2015 with a dedicated project officer and 14 teachers – all volunteers and mainly Unite Community members – trained by the union. After leaflets in four languages (Polish, Russian, Lithuanian, Latvian) were distributed outside the factory by members of Community Unite, 150 workers showed up at the Miners' Welfare Association in Shirebrook to take part in lessons. This was no mean achievement given that workers often walked several miles to work to do a twelve-hour shift before coming to classes. Although a small proportion of the workforce it was a gateway into the migrant community. Brian, a key member of Unite Community active in organising the ESOL classes, explained that:

> They were never an end in themselves (although they have been of great benefit to scores of individual students) but a means to building a bridge between Unite and Sports Direct workers and a forum via which problems could be aired. The ESOL classes were one of the few places where workers could actually meet and discuss their problems.

Unite activists and language teachers used the classes to collect information about working conditions in the warehouse that was fed into the House of Commons Select Committee report (2016). Moreover, the ESOL classes built trust and were used to explain the role of trade unions. Regular contact between the migrant workers and the union meant that they could identify and respond to wider issues that workers were facing, particularly in relation to housing and education. In one case a Romanian family had two additional workers lodged in their house by gangmasters (who brought workers in and supplied them to the agencies) without consultation and against their wishes. The Unite project officer was able to contact the local councillor to intervene.

Fundamental to the success of the campaign was the activity and participation of the Unite Community branch, with members from

Chesterfield, Nottingham and Derby, often retired activists, being the 'boots on the ground'. Mobilising around 30 people at short notice, they were involved in leafleting, delivering ESOL classes and mounting protests outside the factory and in the nearest towns such as Chesterfield. Cementing relationships with warehouse workers was helped by the proximity of key Unite Community members living in, or very near to, Shirebrook in contrast to union offices in Derby and Nottingham that were up to a one-hour drive away.

Two Steps Forward, Two Steps Back?

Given the exposure of the 'gulag' conditions in his warehouse, Mike Ashley was tenacious in denying persistent and widespread criticisms. After initially refusing to appear at the House of Commons Select Committee, he turned up at short notice and appeared 'shocked' at the testament of his workers. As the Select Committee pointed out, his denials were not very convincing given that he was on site at least once a week. Under duress he asked for 'space' to make some changes to and improve working practices in his warehouse (Farrell and Butler, 2016).

There were few improvements and the theme of the demonstration of the supporters' group at the 2017 Annual General Meeting of Sports Direct was 'Business as Usual'. It turned out that the 'space' Mike Ashley needed was to introduce some mechanisation and develop new forms of surveillance and exploitation. Small improvements were made: the 'six strikes and you are out' policy was stopped, a nurse was employed on the premises and there were fewer searches. One well-publicised win was the agreement that Unite secured for workers to receive backdated pay to the tune of £1 million given for unpaid, compulsory queuing. However, although this benefitted some workers as much as £1,000, this was a hollow victory for the 1,700 employed by Transline who lost out when it declared itself bankrupt (Goodley, 2017). These gains were important for the workers, but Mike Ashley reneged on his most important promise to transfer workers from the employment agencies on to to direct contracts. After a few hundred workers, who had started working at Sports Direct between 2006 and 2008, were given direct contracts with the firm, further transfers to permanent contracts dried up completely. It is a case of two steps forward and two back as other aspects of work have deteriorated.

The introduction of mechanisation after 2018 in the form of picking machinery has increased the amount of dust and raised temperatures to dangerously high levels. Workers now receive their orders through headsets; one worker described this as feeling though they were 'an extension of a computer', reduced to its arms and legs.

In 2019 one worker described how they had been forced into another battle when Sports Direct issued an edict that mobile phones had to be left in lockers. This removed a vital line of communication between parents and schools if there was an emergency with their children. The hotline set up by the management to offset workers' complaints was not answered. Pressure from local schools and a petition by Unite Community members outside the factory forced management to back down. At the time of writing a new battle was developing over the use of body cameras worn by supervisors to intimidate workers.

By 2016 a functioning branch of Unite had been established with trained officers and representatives and regular monthly members' meetings, although these were held in the village of Shirebrook and not in the warehouse itself. The membership of the Sports Direct Unite branch at Shirebrook is far from including the majority of workers and there are obstacles to its growth. Although there is a recognition agreement between Unite and management, one union representative described to me how s/he was 'fobbed off' when they tried to raise issues and complaints. Participation in, and the visibility of, the union are highly uneven. Where workers were near to union activists the value of membership was clear: they could get advice and representation in disciplinary hearings. However, in other parts of the warehouse the visibility of the union is much lower. Patchy membership and participation are not unusual in large workplaces, but nevertheless it presents obstacles in terms of recruiting, retaining and mobilising workers. By 2020, after key activists were ill, furloughed or had left the warehouse, the branch became semi-dormant. But Brian, the Unite Community activist, explained that this was far from being the end of the campaign:

> There remains a high level of interest in Sports Direct among the local community branches and there is potential to revive the branch. There is also a need to address conditions in other local workplaces. Sports Direct was only ever the tip of the iceberg. There

are numerous warehouses and distribution centres being set up. Although most don't seem as bad as Sports Direct and some offer better wages, the work is often still precarious and has a culture of bullying, with restrictions on break times and toilet breaks. We have the potential to become a catalyst for organising against these conditions.

A MISSED OPPORTUNITY

In the case of Sports Direct – despite the high-profile publicity in the national press, the publication of a House of Commons Select Committee report and numerous lively protests outside the warehouse and in the locality – the gains seem to be disappointing, both in terms of winning better working conditions and union organisation. It is a contrasting story to the stunning victories achieved by the cleaners at the University of London and BEIS government department that we saw in Chapter 9.

One problem with the campaign against Sports Direct, cited by both Unite Community activists and one of the Unite officials, was a lack of consistency from the union. In 2015 the first big push to systematically forge a relationship with the workers through ESOL resulted in building the membership, and external stunts meant that the campaign had a national profile, but the momentum was lost. The ESOL project only lasted for two years and, according to one Unite Community activist, winding down the campaign on the basis of the empty promises of Mike Ashley 'pulled the rug' from under the campaign and was 'a big mistake'. The campaign was highly dependent on the commitment and diligence of two individuals who worked for Unite. But after the Unite regional official left the region and the two-year contract of the ESOL project worker expired, one Unite representative in the Sports Direct warehouse told me that it signalled the 'end of the good years'. The regional official had been the backbone of supporting activists and consistently provided advice at short notice. Oversight of the Sports Direct branch is now part of the heavy workload of another regional official who is responsible for many other workplaces. Brian, a Community Unite activist, argued that: 'There was never a long-term commitment from Unite. Bernie [Unite project worker] was only ever on short-term projects –

the longest being for six months. She had five contracts in two years. She was frequently just as vulnerable to losing her job at any time as Sport Direct workers.'

The lack of consistency in the campaign and attempts to build an active branch at Sports Direct have to be understood in the wider politics of trade unions and their internal dynamics. Trade unions are not immune from metrics, and the Unite hierarchy wanted to know how resources translated into new membership. A Unite delegate to conference in 2018 told that s/he overheard some officials commenting that recruiting these precarious workers was 'more trouble than it was worth'. Others have criticised the lack of integration of the Sports Direct branch with the wider union machinery of Unite. The Sports Direct warehouse branch is located in a region where there are traditional industrial branches with a skilled and stable workforce and more extensive financial resources. At best they had shown little enthusiasm for the Sports Direct campaign; as one member of Unite Community commented, 'they live in a different world'.

The most glaring factor in the success of the campaign against Sports Direct was the energy, commitment and persistence of individual Unite Community members who provided a link between the officialdom of Unite and workers in Sports Direct. However, while they were the 'boots on the ground' they were no substitute for activism inside the warehouse. Industrial action was never on the agenda. Beverly, a Community Unite activist closely involved with the campaign, argued that although some gains had been made, there is 'only so much that can be done from the outside … we need more members inside Sports Direct to make sure its employees are properly protected'.

Many workers in companies like Amazon are already unionised in Europe. In the US the fight to form a trade union in an Amazon warehouse in Alabama has opened the door to organising in warehouses that are aggressively anti-union. However, in Britain despite the Herculean efforts and persistence of two Unite full-timers the campaign at Sports Direct was a missed opportunity by Unite. A victory against a particularly pernicious employer would have been an inspiration to other workers in the poorly unionised warehouse sector. Laudable and successful though the campaign was in terms of naming and shaming Sports Direct and forging links with workers through ESOL classes, Unite pulled its punches by not using these as a stepping stone for industrial action. As we have seen with the cleaners (and as we shall with teachers and university workers), taking action and campaigning is what builds unions. When workers see that something is happening, they join unions, get involved and gain confidence, and a dispute can trigger this process even if it starts from quite a low base. But Mike Ashley's feet were never really put to the fire. He could ride out verbal criticism, but a walkout or strike would have hit him where it hurt: his pocket and his profits.

11
Education Workers on the Front Line

A NEW PROLETARIAT, NEW ACTIVISM

Hundreds of striking workers on the steps of the iconic Parkinson Building at the University of Leeds on a snowy day in February 2018 – a brazier, dancing, singing and chanting. With union banners and homemade placards this vibrant picket line was replicated up and down the country. Nobody could have predicted that the biggest and most bitter industrial dispute between 2011 and 2020 would have come from university workers – and even more surprising that the eruption of anger was triggered by proposed detrimental changes to pensions. This explosive struggle over pensions was fuelled by accumulated grievances about stagnating pay, casualisation and the intensification of work. Lecturers, professors and academic-related and casualised staff took 36 days of strike action over a period of two years.[1]

In 2020 teachers and workers in schools went into the struggle. They joined the NEU in their tens of thousands and over a thousand young, new representatives came forward. They were angry at the proposed hasty and deadly reopening of schools in June of that year during the Covid-19 pandemic that made them feel like 'lambs to the slaughter'. The mass participation of teachers in Zoom meetings – nationally, locally and in schools – sent an overwhelming message that they would not compromise the health and safety of their members and local people. They forced the government into humiliating retreats over fully reopening schools, both in May 2020 and again in January 2021.

Teachers, further and higher education lecturers and researchers have the shared experience of the proletarianisation of their jobs. Education at every level – from tiny children to adult education – has been marketised and the people who work in it subjected to targets, metrics, micro-management and the intensification of work. Universities have been experiments in and a crucible for neoliberalism, and those that work in schools have been in a fight for their lives and for those in their communities as the government has played fast and loose with their health and safety. There have been similar experiences for those that work in further and adult education where the starvation of funding and constant restructuring have gone largely unnoticed because of the working-class young people and adults that they educate. This chapter looks at the struggles of university lecturers and researchers and how the attack on pensions ignited the touch-paper for action from below. It goes on to look at how angry teachers have fought against the government during the pandemic by linking with their communities and tapping into broader political questions about education and the Black Lives Matter movement. In both cases there was a groundswell of struggle from below that generated a new layer of activists who have transformed branches, regions and national structures.

THE NEOLIBERAL UNIVERSITY: A NEW ARENA FOR STRUGGLE

Working in the Higher Education 'Sausage Factory'

There are few workplaces in the 'public' sector where neoliberal ideas have taken such a grip as in universities (see Chapter 3). In 2018 what began as a seemingly arcane dispute about pensions exploded into a full-blown challenge to marketisation in three waves of strike action between 2018 and 2020.

Higher education had been a relatively good place to work, with reasonable levels of job satisfaction and autonomy – in comparison with other groups of workers at least. At the beginning of my time working in one of these 'citadels of knowledge' in 1991, teaching loads were negotiated and there was a permissive and supportive attitude to research. Classes were small enough to know the names of the students you taught and conversations with colleagues about the content of courses were lively and argumentative but driven by pedagogic concerns and intellectual debate. The expansion of higher education without the resources to support it, and particularly the meteoric rise of business schools, came home to roost. In 2004 I taught a first-year introduction to economics module of 800 students with an army of hourly paid lecturers – a threefold increase in numbers on the previous year. As the fees increased to £9,000 in 2012, managers began to refer to students as customers. We were asked to wear large badges for the first few weeks of term, not dissimilar to those worn by staff in supermarkets, that read 'Here to Help'. By the time I left in 2018 the corridor between my office and the library (or Learning Resource Centre as it was called) was flanked by giant airbrushed photographs of individuals and teams deemed by the top corridor to have gone the extra mile.

Paranoia about losing the 'privilege' of recruiting overseas students fuelled a 'Fortress University' policy. Passport checks were demanded for anyone external 'engaged in university business' – even for academics from other institutions coming to do a seminar for no fee. Overworked colleagues challenged calculations about their teaching loads to the nth degree in a desperate effort to protect themselves from being run into the ground. Critical thinking was treated as a nuisance and a deviation from the latest edicts handed down from management: collegiality and discussion were replaced by one-way talks from managers presenting carefully prepared sets of statistics. These dubious graphs and tables formed the basis of praise and chastisement, with exhortations to improve our performance underpinned by veiled references to job security if we failed to do so. For lecturers and researchers, being at the receiving end of micro-management and metrics has cemented their experience of the proletarianisation of work in universities.

Marketisation, Metrics and Micro-Management

In 1962 Britain only had 31 universities educating 118,000 students – just 4 per cent of school leavers – the vast majority of whom were from middle-class backgrounds (Vernon, 2018). By 2018, 2.3 million students were studying at higher education institutions in the UK that employed 430,000 workers (Universities UK, 2019). Universities have transformed beyond all recognition to become corporate entities and global players that contributed 1.2 per cent to the GDP of the UK in 2019 (ibid.). The drift to privatisation was consolidated by the increase in fees in 2012; this entailed shifting the burden of funding universities from the state to students (whose fees now make up 50 per cent of university income). The final ingredient for completing the marketisation of universities was the Higher Education and Research Act (2017) that opened the door to competition from the private sector which was given degree-awarding powers. The gateway was opened for global for-profit firms, such as Kaplan and BPP, and private equity to compete with existing universities (that have charity status) by extracting profits through state-supported loans.

The metrics of the Research Excellence Framework (REF) and the Teaching Excellence Framework (TEF), intrinsic to marketisation, have brought intense and cut-throat competition between higher education institutions, and embedded stress and bullying for university workers. First introduced in 1989, the REF (and its previous iterations) purports to measure the quality of research output in four-year cycles – the results of which determine future research funding. The implications of this for academics have been draconian, with staff transferred to dead-end teaching-only contracts (note the undervaluation of teaching), professors eased into early retirement and swathes of younger academics working for years on short-term contracts. One effect has been to produce a syndrome of stars and minions where professors deemed to be research stars are poached from other institutions, lured by non-disclosed salaries way off the pay scales, often without setting foot in the university that payrolls them. These salaries are effectively subsidised by an army of poorly paid lecturers on fixed-term contracts. Casualisation is a massive problem in higher education with the UCU (2019a) estimating that this 'reserve army' of labour is doing between 25 and 30 per cent of teaching in many universities. Less visible are the 70 per cent of 49,000 researchers employed in the sector, who were on fixed-term contracts in 2019. The TEF uses the National Student Survey, based on a series of metrics (student satisfaction, retention, employability and 'learning gain'), to generate a league table of universities. The TEF is deeply unpopular with lecturers and has led to the micro-management and constant surveillance of teaching on the basis of highly spurious statistics. The marketisation and privatisation of higher education is the context in which a new union, the UCU, was born and later provided the tinder for one of the biggest strikes in the second decade of the millennium.

The Birth of the Universities and Colleges Union: 2006

The birth of the UCU in June 2006, from a merger between the Association of University Teachers (AUT) and the National Association of Teachers in Further and Higher Education (NATFHE), heralded a new era of trade unionism on university campuses.[2] With 120,000 members at its formation, the UCU was the largest further and higher education union in the world. The commitment of the new union to equality issues, fighting privatisation in the sector and challenging mass casualisation are reflected in excellent research materials, comprehensive policies and robust press releases. There have been some exemplary campaigns waged by the UCU nationally against redundancies and union derecognition when the full weight of the union has been thrown behind battles in individual universities. A handful of one-day strikes were called between 2006 and 2018 to support pay campaigns. In terms of the wider politics of the UCU, there are a string of affiliations to progressive campaigns in Britain and internationally.

However, as we saw in Chapter 6, there are two competing visions of unions. In the UCU this cannot be reduced to, but is most starkly illustrated by, a historic division since 2006. On the one hand there are sections of the bureaucracy, supported by elected members of a group that calls itself the Independent Broad Left (a misnomer because its leading lights are members of the Communist Party and fellow travellers), that have been a conservative force, veering towards servicing and pessimistic about the willingness of members to take action. On the other hand, the UCU Left (founded in 2006, see Chapter 6, comprising individual activists, left-wing Labour Party members and members of the Socialist Workers Party) describes itself as, 'committed to building a member-led, democratic UCU, one which is founded on a strategy of campaigning and collective action rather than one based solely on servicing members and individual casework'.[3] But in February 2018, when British universities became engulfed in the largest dispute in their history, a wave of strikes threw up a new layer of grassroots activists that also put the issue of democracy in the UCU centre stage.

An Explosion of Anger and Action: The 2018 Wave of Strikes

Triggered by the unlikely catalyst of changes to pensions, lecturers, professors, researchers and administrative staff took 14 days of strike action in February and March 2018. Simply put, the University Superannuation Scheme (USS) wanted to end 'defined benefits', meaning that pensions would have been tied to the vagaries of the stock market. This draconian proposal would make university workers lose £10,000 per year of their pensions on average, with women and ethnic minority staff even worse off because of the gender and equality gap. The kernel of the USS case was premised on the risk calculation made by the consultancy firm PricewaterhouseCoopers, which argued that the scheme was running a deficit – a 'fact' vigorously challenged by the UCU and in even more forensic detail by the UCU Left and an emerging group of activists – USS Briefs.[4]

In 2018 the voter turnout in pre-92 universities balloted in response to the employers' offensive was 59.8 per cent with 88.1 per cent in favour of strike action. Not only the employers but even left-wing activists in the union hugely underestimated the level of anger that this apparently abstruse issue triggered. Activist Mark Bergfeld describes his surprise at the scale of and enthusiasm for the strikes:

> Those like myself who had been pessimistic about organising a strike were proven wrong and suddenly found themselves attaining creative, large and vibrant picket lines ... As a sense of collectivism emerged in action on the picket lines, discussion with colleagues completely changed the tone ... On the picket line, the walls between casualised staff and professors, different academic fields and the isolation of academic labour were increasingly being broken down day by day. At Queen Mary University of London, union members who had not previously attended branch meetings came out and even started taking leading roles. At the national level it was reported that 3,000 employees had joined the union ahead of the strikes, with more than 2,000 applications processed in the first week alone. As the strike unfolded the aesthetics and tactics of the strike came to resemble the 2010 student movement. (2018: 234)

This was the experience of the strikes up and down the country. Sean Wallis, branch president of University College London, the UCU's largest branch, describes how 200 people turned up to the picket line on the first day, most of whom had never been on one before and with an estimated 10 per cent who were not even union members at the time (see Figure 11.1). Membership grew by 50 per cent in a couple of weeks. Anne Alexander, a committee member from Cambridge University UCU, told me how, in the weeks after the strike days were announced in 2018, membership doubled from 700 to 1,400. The numbers kept growing as new waves of strikes were announced and by April 2020 the branch had 2,000 members – an increase of 180 per cent in two years.

Figure 11.1 University workers at Kings College London picket during strike to defend pensions in February 2018
Source: Guy Smallman.

A small glimmer of this vibrancy and the thousands that flooded to participate on the first day can be seen in the music videos of strikers on snowy picket lines at the universities of Leeds (University of Leeds UCU, 2018) and Cambridge (Cambridge UCU, 2018). New forms of democracy emerged, with strike committees supplanting or transforming existing branch committees and the locus for decision making shifting to mass meetings. Sean Wallis explained the importance of regular mass meetings:

> The advantage of the strike meeting became obvious: with over a hundred in the room every day, we could decide to picket negotiation meetings or march on UCEA [the employers' body]. We could debate potential offers from the employers, elect delegates to go to meetings. And we organised teach-outs to involve student supporters ... union branches relearned the democracy and power of the mass strike meeting. This involved democracy, not just in the traditional sense of decision making, but also accountability. Reps were elected at meetings and reported back on other meetings ... Local strike committees and similar open grassroots assemblies had a dual role, initially to maintain the strike and take local initiatives, and then as the dispute developed, to protect the interest of members in the strike from the union bureaucracy and National Executive Committee holding back.

An unofficial Facebook group was set up with more than 1,000 members to share information, exchange experiences and plan tactics and strategies. WhatsApp strike groups in individual institutions kept people informed about developments on a day-to-day and hourly basis. The media were quick to interview students claiming their education was being disrupted, but the reality was massive support from students, with an estimated 200,000 signing a petition in support. By the third week students across the country in more than 15 institutions occupied the office of vice chancellors and university buildings. International Women's Day coincided with the UCU strike and activists organised teach-outs on the gender pay gap, #metoo and the history of women's strikes. Held off university campuses, these teach-outs sprang up everywhere, aiming to repoliticise the neoliberal university with discussions on climate change, casualisation and culture.

However, the enthusiasm and expectations of the rank and file collided with the official union, its structure and its rules. Such was the engagement and resolution of grassroots members that an attempt by the UCU leadership to agree a hugely detrimental deal in March 2018 was met with fury by members. Branches were given less than 24 hours to consider the deal. It was roundly rejected. The hashtag

#nocapitulation spread like wildfire and a petition launched by the Facebook group collected 5,000 signatures in an hour. London universities held early-morning meetings and enraged members congregated outside union headquarters to lobby delegates attending the Higher Education Committee, who rejected the deal. This was a huge rank-and-file demonstration organised rapidly via networks of activists across London. The vitriol spilled over to the UCU conference in May 2018. A motion critical of the then general secretary, Sally Hunt, holding her accountable, was ruled out of order. Some staff employed by the UCU took the extraordinary step of closing the conference down twice, on the grounds that a member of their union Unite was being attacked.

The genie was out of the bottle. A new layer of activists mobilised with the outcome that in the election for general secretary in 2019, following Sally Hunt's resignation on grounds of ill health, a young, relatively unknown candidate, Jo Grady, was elected. She had been invigorated by the 2018 strikes and was at the fore of the USS Briefs group that had emerged during the action. The 49 per cent of votes that she garnered, combined with the 25 per cent vote for the UCU Left candidate, meant that on the highest turnout ever for a general secretary election (20 per cent), the left won 74 per cent of the votes, a reflection of the anger of activists. This result shook union officialdom to its core as the UCU full-timer candidate, viewed as the natural successor to Sally Hunt, received only 26 per cent of the vote.

Grassroots Activism

At the UCU Congress in May 2019 delegates voted to run a broader 'Four Fights' campaign alongside the dispute to defend pensions. The inclusion of demands for equality, and an end to casualisation and high workloads along with increasing pay, inspired a new round of action that included post-92 universities alongside colleagues in pre-92 universities. Sixty institutions got over the 50 per cent turnout threshold and eight days of strikes were held in November and December of 2019. As the campaign gathered pace, another 14 universities reballoted and voted to support 14 days of strikes in February and March 2020. The strikes of 2018 to 2020 galvanised not only existing activists and members of the UCU Left, but created a new layer in the pre-92 sector who wanted to get involved in the UCU in the workplace and at regional and national level. For example, a new network of rank-and-file members grew up around USS Briefs, which had championed the candidacy of Jo Grady.[5]

Departmental reps, committee members, branch officers and rank-and-file activists in universities threw themselves into building the strikes on the ground. One new UCU activist from a northern university, Mark, cited examples of universities where new activists had challenged moribund and 'dysfunctional' branch committees in the pre-92 sector to win a majority of the places. Branches were (re)invigorated by the involvement of a wave of established and new militants. He described how their union committee established a network of representatives that covered most of the university, ensured an accurate membership list and 'used resources to plug any gaps'. Phone banks and 'walk-throughs' were used to give updates and mobilise their members, which was reflected in hundreds attending large branch meetings with turnouts of around 60 per cent of the membership. Mark explained that activity and action, at the centre of recruitment and building branches, was set in the wider context of and connected to campaigns against climate change, racism, transphobia and defending migrant workers. The mass assemblies, strike committees, teach-outs and vibrant demonstrations could not be further away from the gentleman's club of the AUT that dominated many campuses in old universities before 2006. Anne described this generation of activists as 'a new cadre in the union forged in successive strike action'.

The Covid-19 Pandemic: A New Wave of Struggle

The Covid-19 pandemic presented a sharp challenge to UCU activists. The high-risk expansion and indebtedness of the higher education sector was exposed, with the Institute for Fiscal Studies estimating losses in the sector ranging from £3 billion to £19 billion (Drayton and Waltmann, 2020). However, these were unevenly spread between institutions and 13 higher education institutions were highlighted as being in danger of bankruptcy. The response of managers to this dire situation of their own making was a determination to get students onto campus, paying their fees and high rents with no regard for the 'student experience' and their physical and mental well-being. The outcome was soaring rates of Covid-19 across campuses in October 2020. The health and safety of university lecturers was compromised as some managers demanded that they engage in face-to-face teaching. Stress levels were ramped up as work intensified when lecturers were expected to deliver online teaching, often without support, at short notice. Carlo, a longstanding member of the UCU National Executive Committee, explained how the campaign to increase wages, defend pensions, fight for equality and oppose casualisation were put on the back burner:

> Once the Covid-19 pandemic hit in the spring of 2020 the 'Four Fights' campaign simply fizzled out. Our members were hit with enormous workloads as teaching went online and the momentum was lost as it was difficult to mobilise members over the summer. This was reflected in only 30.4 per cent of members voting in the ballot for action in July – that didn't hit the legal threshold.

The national democratic structures of the UCU were slow to respond to the Covid-19 crisis, but employers were not. Citing catastrophic losses, exaggerated in the case of some institutions, university managements were quick to go on the offensive against their workers, threatening redundancies and the non-renewal of fixed-term contracts. Carlo went on to argue that:

> It was activists from below who organised the fightback. Four branches – Imperial College, the universities of Liverpool and Reading and SOAS – who were in a struggle with their managements over jobs, came together to form the UCU Solidarity Movement in May 2020. Since then it has mushroomed to include other branches, including further education colleges, going into struggle over redundancies and national solidarity rallies have been held.

The website of the UCU Solidarity Movement urges the UCU to 'up its game' and use models of mass participation to shape campaigning and industrial strategy to build a united national resistance to defend jobs rather than leaving it to individual branches.

In September/October 2020 the UCU nationally responded by holding a six-part organising school with Jane McAlevey. In the

2020 autumn term branches went into action and there were successful ballots for strike action over health and safety and to get all teaching moved online, while the University of Brighton held a strike over redundancies. By January 2021 Jo Grady, general secretary of the UCU, announced that bringing students back onto campuses early and dangerous health and safety practices such as face-to-face teaching would be met with strike ballots. The next section discusses how the NEU and its rank-and-file members were quick off the running blocks in organising and forcing the government into two retreats.

TEACHERS AND EDUCATION WORKERS FIGHTING FOR THEIR LIVES AND COMMUNITIES

The Covid-19 pandemic was a huge test for the labour movement, one that many trade union leaders failed miserably as they were sucked into the orbit of the government's rhetoric of national unity. Some even closed down democratic spaces. The leadership of the NEU and its officialdom set itself apart by being much bolder and combative both in their public stance and in galvanising and supporting their membership. There were three key moments in the struggle: in March as the pandemic broke out; in May when the government announced fully reopening the schools; and in late December/early January when the prime minister, Boris Johnson, insisted that schools would be fully open despite the pandemic raging out of control. In the early stages of the pandemic in March 2020, when battles were taking off in schools over getting vulnerable staff sent home, Kevin Courtney, joint general secretary of the NEU, posted a robust message on protecting vulnerable workers on an unofficial union Facebook page: 'If your head says they [vulnerable workers] have to be in, tell them no. Tell them it's your union's advice. Tell them you will work from home. Tell them you will see them in court. Tell them if they mess you about there will be trouble' (Thomas, 2020: 45).

Desperate to keep profits flowing, and blind to the lessons of other countries, the government was slow to close down the economy. Its hand was eventually forced as death rates from Covid-19 soared and on 18 March they announced the closure of schools at two days' notice. The work of the NEU was focused on the safety of its members who were in school teaching vulnerable children and those of key workers, and on ensuring that those at risk worked from home. In parallel the NEU began the mass training of reps in health and safety.

Getting Organised: Spring 2020

In spring 2020 a major struggle erupted over the bombshell announcement from the government on 11 May 2020, without any consultation with the teaching unions or even head teachers, that schools would reopen on 1 June. This sent a wave of shock and fear through educators. One district secretary, Paul, explained: 'Health and safety concerns suddenly became a matter of life and death'; another, Sobia, said that teachers and support workers felt 'like lambs to the slaughter basically exposing us to a deadly situation'.

Rank-and-file activists and the NEU nationally went into action to organise using the protection encapsulated in Section 44 of the Employment Rights Act 1996 that entitles an employee to withdraw from and refuse to return to an unsafe workplace. Twenty thousand new members signed up to the NEU in the first few weeks. According to Claire, a young NEU rep in an inner London school, the government's announcement was like lighting the blue touchpaper. Her school immediately elected a young black woman as health and safety rep and the 40 teaching assistants in the GMB trade union elected a rep for the first time. She explained how the NEU and GMB in the school acted as if they were one union and quickly collected 54 signatures on a letter to the head saying that they would not take part in planning for a return to work until the 'five [safety] tests' set by the NEU nationally were met. The spontaneous bottom-up response in this school was replicated up and down the country: the number of reps recruited nationally leapt from 10,000 to 13,000 – an increase of nearly a third.

Elected district secretaries of the NEU had been a relatively conservative layer, often taking a 'top-down' servicing approach: many had been absorbed into the union machinery. An anonymous source from the NEU described a planned programme that was nicknamed 'Operation Asteroid' because it aimed at replacing time-serving dinosaurs in the union by training new activists who could invigorate district organisation. Jon Hegerty, an NEU training officer, describes the union's secretaries training: 'Although statistics collected in 2019 showed that 67 district secretaries were over 50 and 58 per cent were male – despite men comprising only 24 per cent of membership – these figures are changing. We are recruiting and training new secretaries who more reflect the membership.'

The picture was uneven; while some districts had no tradition of organising, merely keeping the union machinery ticking over, others played a leading role in recruiting reps, involving members and coordinating action. Sobia, the joint secretary of a Greater London district, told me: 'We were very strategic ... bloody hell we worked day and night – but it was worth it. We had spreadsheet after spreadsheet so we could be methodical and target schools without reps. We made individual calls to members and then held Zoom meetings for members in those schools.'

Paul, a district secretary in the Midlands, explained that they were able to 'hit the ground running' by building on a base of reps recruited in the previous year as part of a campaign they had won over inconsistencies in pay scales that had short-changed their members. He outlined how they built action and increased the infrastructure of reps further:

> We built the campaign school by school. The committee divided schools into three categories. Where we knew there were established and active reps, we encouraged them to have members' meetings, provided a script and offered that they could shadow me in meetings with other schools. In schools with new reps, who were less confident, we offered to help with organising school meetings. Where there was no rep we set up Zoom meetings and got reps elected.

One of the biggest gains was the involvement of primary school teachers who had been 'historically difficult' to involve in the NEU (and previously the NUT). As Sobia put it: 'The big issue was in primary schools as the government wanted them to be crèches. It had been more difficult to recruit reps in primary schools because the workload is a killer. But teachers were being put in dangerous situations and they stepped up.'

Using the Organising Agenda

The framework for this activity was provided by the robust messages, activity and support of the NEU. As Sobia explained: 'the strong and clear lead from [the] national union gave us confidence to push back … and it empowered us'. Rather than closing down democratic spaces, as had been done by other unions, the NEU used its return to the workplace organising agenda to build the union and start laying down the conditions for the safe opening of schools. National virtual 'town hall' meetings were held on Zoom attended by thousands of members and activists. At NEU central office a call hub was set up to start ringing around to build contact with reps and gauge what was happening in schools: Jon Hegerty from the organising department estimated that in the first three weeks they contacted over a third of reps. In total 650 representatives were trained in health and safety using Zoom (both specialist reps and general school reps). The 'organising' message was consolidated with 'inspirational' Zoom seminars with Jane McAlevey, author of *No Shortcuts* (2016) (see discussion in Chapter 6). Militancy and momentum from rank-and-file teachers and the NEU nationally resulted in a huge victory in pushing back the date for the opening of schools and stipulating the conditions under which teachers would return. On 1 June only 30 per cent of schools opened, with an average of 50 per cent of children returning. On 9 June the government beat a humiliating retreat, conceding that it was not feasible to open schools and getting children back would be postponed to September.

The Importance of Politics and Community

A community-based approach was used to support workplace organisation. Jon Hegerty explained that:

> Over the last six or seven years, and inspired by the Chicago teachers strike in 2012, we have shifted to a social-movement model. Concretely this means working with parents – we have strong links with the School Funding Campaign and Anti-Academies Alliance … working with Black Lives Matter has allowed us to build a really strong anti-racist campaign in the union … we held a Zoom meeting with US civil rights veteran Jesse Jackson, MP Diane Abbott and a young, black woman representative from the NEU.

In her district Sobia explained how they linked up with the community:

> One of the biggest concerns was to keep parents on board – especially in the light of 'teacher bashing' by the right-wing press. Our message of 'our local community versus the government' went down really well. We were able to build on our parent network from the anti-academy campaign and we managed to get nearly 600 people to a Zoom meeting organised by the campaign and the NEU. Even local MPs and the leader of the council came and were really surprised by the support – they must have thought 'bloody hell we had better listen to these teachers' and they made all sorts of promises and talked about valuing teachers.

National NEU support for the Black Lives Matter movement opened up space for linking anti-racism and the dispute at a local level. When George Floyd was murdered by the police in Minneapolis on 20 May 2020, the Black Lives Matter movement reignited, first in the US then quickly reverberated round the world. In Claire's school the head initially refused the request of the black health and safety rep to make a statement in support of the Black Lives Matter campaign. Claire explained that:

> [L]earning from the model that had been used to prevent the reopening of the school the health and safety rep drafted a letter signed by over half the educators in the school to visibly support the movement and the head teacher relented. The fact that a young black teacher had taken a lead role in fighting for the safety of members put a handful of NEU and GMB members who argued that 'all lives matter' in a contradictory position and meant that they had to engage with the arguments. Our fight for health and safety went hand in hand with heightened political awareness and debates on racism and wider educational issues.

Into Battle Again: December 2020/January 2021

In early September 2020 Covid-19 cases were low and there was no argument for not fully reopening the schools. Enormous efforts were made by teachers and head teachers to keep schools as safe as they could be with the creation of complex bubble arrangements: one small inner London primary school was spending £100 of its budget a month on hand sanitiser. This was the lull before the storm. As the number of Covid-19 cases crept up there were increasing demands for a circuit breaker in the spread of the virus around the October half term. But by mid-December, as the number of cases reported and deaths soared, there was a sharp shift in feeling among educators that the schools needed to close. The government's announcement that teachers would be expected to deliver mass testing in schools using lateral flow tests, which were deemed to be unreliable, alienated their core constituency of head teachers and set the stage for another confrontation with the government. Militant associations, rank-and-file teachers and the NEU went into action again. Sobia explained:

> On 30 December several London associations came together and started agitating to get schools closed. We held a Zoom meeting, not expecting a huge attendance as it was the middle of the holidays. Five hundred teachers attended – the maximum that Zoom allowed. I think this played a massive role in creating the mood and momentum for pushing back against the full reopening of schools after Xmas. It gave the leadership of the union confidence – they can't be ten steps ahead of their members – it was a massive signal about how worried our members were. We were really impressed with how the leadership responded.

An emergency meeting of the National Executive Committee was held on Saturday 2 January; later in the day there was a meeting for 4,000 school reps. The following day – Sunday – a national meeting was live streamed with 500,000 teachers watching at any one time. On Sunday 3 January the prime minister, Boris Johnson, was still insisting that schools would reopen fully, but after the massive explosion of protest and pressure from teachers he performed yet another spectacular U-turn and the following day announced that schools would not be fully open until further notice.

Revitalising the Union

These struggles changed what the NEU looked like. The disputes tapped into a new group of lower-paid educators – mainly teaching assistants. Historically organised by the GMB and Unison, both glacially slow to take the lead, 30 per cent of new recruits to the NEU were teaching assistants. But the stark outcome was the changing demographics of new reps. Jon Hegerty cited statistics showing a massive increase in the number of young reps getting involved: the percentage of reps under 40 increased from 31 per cent before the Covid-19 pandemic to 50 per cent in summer 2020 (six months after lockdown): there was a step change in black reps (3 per cent to 7 per cent), women reps (69 per cent to 73 per cent) and LGBT reps (2 per cent to 5 per cent) in the same period. The disproportionate impact of Covid-19 on black and Asian communities and widespread protests by Black Lives Matter in Britain encouraged more educators from BAME backgrounds to come forward and get involved. According to Sobia, 'The reps are often young women and many of them are black. It really helps that I am from that background – you can't be what you can't see.'

The collectivism and confidence engendered by winning victories in individual schools and forcing a humiliating climbdown by the government empowered and energised a new layer of activists. It opened up a space for thinking about wider issues of pedagogy and the curriculum. Claire argued that: 'Head teachers momentarily lost control – that enabled teachers to think about holistic and child-centred approaches. It resurrected the big questions about testing children, cheating [with the tests] and class size.'

These victories had inspired confidence in members to fight on other issues. Sobia explained: 'The energy we put in forging connections will pay dividends in the long run. We had started from a dismal position of managerial micro-management and ever present Ofsted and this turned union organisation around – you won't be able to put these new activists back in the box!'

THIS IS WHAT THE WORKING CLASS LOOKS LIKE

Far from being privileged workers hermetically sealed in the ivory towers of academia, workers in further and higher education have been proletarianised by the neoliberalism of education. The pensions dispute was the lightning rod for grievances over inequality, workload, casualisation and stagnating pay and new activists were born out of this struggle. During the 2020–1 pandemic the fight for health and safety and against aggressive managements making people redundant has spawned another layer of activists and a fresh wave of rank-and-file activity and organisation in universities and colleges. In the face of the pandemic the NEU and rank-and-file activists hit the ground running as schools were put on the front line of defending the safety of their members and communities. Both the UCU and NEU have been transformed by existing and new activists forged in these struggles. This is what the working class looks like in the new millennium. Chapter 12 goes on to look at activists in new areas of the economy.

12
New Kids on the Block

ACTIVISTS IN NEW AND OLD SPACES AND PLACES

Between 2015 and 2020 McDonald's workers walked out three times to join global protests in the fast-food industry; a games developer was sacked for starting a trade union; and a flashmob of couriers occupied the marble-clad lobbies of powerful global companies. This is the changing face of trade unions in the new millennium. The focus of this chapter is on the organisation and self-organisation of workers in places where unions did not previously exist, ranging from games developers at the technological frontier of work to traditional areas of the economy such as 'hospitality' and couriers. Despite these workers being deemed to be unorganisable they have collectivised to win important gains in pay and conditions. The limits and boundaries of trade unions are explored by looking at sex worker organisation – a controversial debate in the British labour movement but one where the tide is turning towards support for decriminalisation and workplace organisation.

THE GAMES INDUSTRY: FIGHTING EXPLOITATION AT THE 'CUTTING EDGE'

In 2019, with a colossal estimated worth of £5.7 billion, media hype and political rhetoric have pointed to the UK's gaming industry as a great success story and exemplar of the creative sector (BBC, 2019). In 2016, 20,430 jobs were directly generated by the games industry, with a further 13,840 in development, publishing and digital and physical retail (UKIE, 2018).[1] With its image of bohemian, autonomous workers and unconventional workplaces the games development industry has all the ingredients of being 'cool' and is lauded for providing liberating and flexible work. This convenient rhetoric of self-made, heroic and plucky individuals diverts attention from domination of the sector by the giant multinational firms who have invested in development studies and publishing networks.[2] Although the industry has generated huge profits for multinational firms and some individuals, ordinary workers in the industry have not shared in these spoils; the dark side is the huge amounts of unpaid overtime expected by employers, with burnout at epidemic levels. With the predominance of young, white and male workers and women's experience of discrimination and sexual harassment, gender politics are more akin to the 1950s than the new millennium.

The 'demanding' working conditions experienced by workers in games development have to be set in the context of an industry that is turbulent and intensively competitive. Major and dynamic technological changes, such as the distribution of smartphone handsets, the affordability of mobile data plans and the arrival of tablet computers, have accelerated the impetus to develop new types of games. The launch of the iPhone in 2007 was quite literally a game changer that led to mobile gaming becoming the dominant player in the industry. It provoked a frenzy of acquisitions as large multinationals cannibalised smaller, profitable firms to tap into a stream of new ideas and innovation.

In contrast to the image of the solitary genius, the process of developing a game is complex and involves working in teams, which in large firms may comprise between 600 and 1,000 people in a physical workplace. Interviewee Raj explained, 'there are conversations – bouncing ideas – lots of communication'. Many positions remain vacant for six to twelve months – hiring workers is expensive and firms need to invest in them. Although, in 2014, 86 per cent of the workforce had permanent contracts, experts in graphics or audio are recruited on fixed-term contracts to do specialist work on a specific game. While this group of workers may be able to command a high fee, in contrast workers in the bottom tier involved in 'quality assurance', that is, testing games before they come on the market, are highly casualised on fixed-term contracts and in some cases this task is outsourced to low-wage countries.

'Crunching': Excessive Overtime

Time to market is a key ingredient of competition in the games sector. Intense pressure to finalise the game, iron out glitches and make last-minute changes demanded by employers is transferred to the workers in the form of expectations of systematic overtime. A survey by the Broadcasting, Entertainment, Communications and Theatre Union (BECTU, 2019) found evidence of 'crunch working' – that is, clocking up 80 hours a week, with regular demands to work 12 to 16 hours a day and through weekends for the six weeks prior to the release of games. Daniel talked about the way in which a spirit of camaraderie emerged from a feeling of 'being in the trenches together' or 'shared trauma'. But work satisfaction has its limits and a BECTU survey revealed that 57 per cent of workers reported experiencing overt bullying and harassment. Steve explained that, 'when we raised concerns about putting in the extra hours the boss suggested that we were in the "wrong job" because we lacked the commitment to go the "extra mile". Sometimes the pressure put on us was more subtle, for example when food was ordered in to discourage us from leaving the office.'

The BECTU survey reported that 53 per cent of workers said that long hours had a negative effect on their personal or home life. This was confirmed by all the interviewees, who described how the pressure of work, particularly at the end of projects, had a highly deleterious effect on their physical and mental health and that of their co-workers. Steve told me that, 'burnout and exhaustion are common, as is a reliance on medication or other substances to get through a project. Long hours of work put pressure on relationships with partners, family and friends ... sitting at a computer for a minimum of eight hours results in serious physical problems.'

GETTING ORGANISED

From a trade union perspective, games developers are a blank page with no recognition agreements and no branches. But there are two parallel initiatives that are gathering pace to organise workers in the industry – from BECTU and the IWGB. Games Workers Unite sprang from a 'flashpoint moment' at the game developers conference in San Francisco in 2018 where there was a sharp debate that asked, 'Are unions good for the industry?' (Dealessandri, 2019; Statt, 2018). In August 2018 there was the first meeting of Games Workers Unite in Britain where attendees took the collective decision to become part of the IWGB. Their campaigning demands are: an end to excessive (unpaid) overtime; an increase in diversity and inclusion; to educate and support workers; and a secure, steady and fair wage. Pay is a big issue; with no national or local framework for scales, what workers are paid is arbitrary within and between firms.

In a situation where even the larger companies have no workplace representation, the general secretary of the games workers' branch of the IWGB described organising as 'flying by the seat of their pants'. He recounts how, when the games workers' branch went public, they received an 'astounding response' from all over Britain and by 2020 had managed to establish regional branches in London (and the South East), the East and West Midlands and the North East. Daisy, a games developer in the Midlands, told me she went to a well-attended meeting organised by the IWGB and walked out having agreed to be the secretary of the new branch. Her enthusiasm came through when she talked about the local monthly members' meetings held in pubs and keeping up with representatives in other regions with a weekly national conference call.

The growing number of workers joining BECTU in Scotland is similarly a bottom-up initiative. According to Fay, a full-time officer of BECTU, 'workers were not only phoning up to join a union, they were asking what they could do at work to get more security and protection ... they were all young and wanted to make a difference'. She describes obstacles to organising in some places where there was 'a culture of competing to show dedication and judging everyone if they left work early. Also workers are cautious about going into battle with their bosses because of a culture of fear from seeing some of the big firms employ anti-union consultants.'

NOT SO COOL FOR WOMEN

While the technology used in games development is cutting edge, women (and men) are having to fight battles that have been, at least partly, won in other sectors. The statistics paint a dismal picture of gender inequality where only 14 per cent of the workforce is female and unequal pay is epidemic. Women are underrepresented in technical roles such as development (9 per cent) and even less in sound and music (5 per cent) and creative development (5 per cent). Minority ethnic workers are even more woefully underrepresented with only 4 per cent in the industry workforce (Ramanan, 2017).

The issue of equal pay looms very large. In 2019 the UK games industry pay gap between men and women widened by 3.5 per cent. Well above the national average in terms of inequality, it was even as high as 64 per cent in one of the largest companies (Taylor, 2019). Helen explained that:

> It was almost impossible to raise the issue of equal pay because of a complete lack of transparency in pay structures. I found out by accident that my male counterpart, in the job for less time than me, was paid 25 per cent more. These gender pay gaps are made worse by secrecy around bonuses that are awarded on an arbitrary basis – in extreme examples these could make up 50 per cent of someone's salary.

Women interviewees reported feeling that they were under pressure to 'prove themselves' and were often at the receiving end of hostile comments about diversity practices. The lack of procedures and policies in games industry firms make women particularly vulnerable to sexual harassment. Daisy reported that challenging sexist comments made by co-workers were often met with accusations of being oversensitive or having no sense of humour – a much-used tactic to undermine and close down women's objections to inappropriate or offensive language or comments. Sexual harassment was rife in the sector, with no procedures or policies to deal with it. It is no surprise then that it was young women in the games industry who were the first to approach BECTU. Along with the issue of pay inequality they wanted to address a culture of long hours where hanging around after work was the norm – further down the line they were thinking about how incompatible this poor work–life balance was with having children or, as one interviewee commented, 'wanting to have a life at all'.

However, moves to collectivise and take up grievances have not been welcomed by all employers, some of whom seem to be affronted that, despite providing pizza and pool tables, their workers are not satisfied with their working conditions. This was highlighted by the sacking of the lead organiser of the new branch of the IWGB, Games Workers of the UK (Quinn, 2019b). Austin Kelmore was working as a senior developer for Ustwo and had been one of three core developers on *Assemble with Care*, one of the major releases by this Bafta award-winning studio. Internal emails claimed that Austin was 'spending too much time on diversity programmes' and 'always putting leadership on the spot' (ibid.). Despite the gains made in winning over (often young) workers to trade union membership and organisation, hostile and aggressive employers mean that the stakes can be high for activists putting their head above the parapet.

NO 'NO-GO' AREAS FOR TRADE UNIONS

A feature of the landscape of struggle since 2015 has been that both mainstream and new trade unions have started organising workers in jobs and sectors that have been largely untouched or neglected by labour organisations. This has focused on (often young) workers in highly precarious work offered by the so-called hospitality sector, couriers and sex workers.

Inhospitable Workplaces in the 'Hospitality' Sector

In 2018 there was a wave of protests across Britain inspired by those taking place globally in the fast-food sector. The small BFAWU (the Bakers' Union for short) organised walkouts in McDonald's and Wetherspoons, while the giant union Unite was behind strikes in a number of restaurants in the TGI chain. Many young, unorganised workers such as fast-food couriers, not members of unions but coordinated by the IWGB and IWW, joined in spontaneously; some Uber Eats couriers and drivers occupied Uber's headquarters.

Few firms epitomise the neoliberal era and the growth of insecure work more than the ubiquitous McDonald's – the world's second largest employer with 1.7 million workers. McDonald's used insecure contracts extensively when it arrived in Britain in 1974 and transferred the business model of the Ford car factory to the fast-food industry. The product is standardised and tasks are fragmented and deskilled in order to 'scientifically manage' every step of the production process with strong managerial control. To maintain this 'well-oiled' machine workers need little training and are replaceable cogs, organised on a just-in-time basis. It is hardly surprising then that recurrent themes reported to Chris, a Bakers' Union organiser, by McDonald's workers are stress and intense work, bullying and arbitrary management and long shifts with short breaks and poor money. By 2019, emboldened by the #metoo movement, the union was flooded with dozens of women employed by McDonald's bringing forward complaints of sexual harassment. In April 2017, in order to trumpet their offer of giving all of its 115,000 UK employees the opportunity to switch to fixed hours, McDonald's did an about-turn and ditched the rhetoric that their workers 'loved the flexibility' of ZHCs. However, despite the fanfare with which this announcement was made, Chris described it as a 'confidence trick' because any offer came at the discretion of local managers and with stringent eligibility conditions. This empty concession did little to quash the appetite for protest among some of its workers.

On mostly insecure contracts, and comprising young people, migrant workers and older people with no other employment opportunities, this group of workers was deemed to be unorganisable. Yet by 2020 there had been three strikes (see Figure 12.1). In September 2017 history was made when workers from two branches of the Bakers' Union – Cambridge and Crayford, South London – went on strike. As a result of the strike, in January 2018 McDonald's gave workers an above-inflation pay rise. On 12 November 2019 members of the Bakers' Union in six London branches went on strike as part of an international day of action for fast-food workers' rights. The demands were for a 'new deal' including £15 per hour, an end to lower youth rates of pay, the choice of guaranteed hours a week, notice of shifts four weeks in advance and union recognition.

There are other success stories in the hospitality sector. In a broad-based initiative the Better than Zero (BTZ) campaign was embedded in large mainstream unions. In 2015, initiated by bar workers who were members of Unite in Scotland, it was given the stamp of approval by the Scottish TUC in 2015. Rather than the campaign being the property of one union, it spans Unite, the PCS, BECTU and the Bakers' Union, as well as including grassroots activists. Since its formation BTZ has had some resounding wins. Unite organiser (and then coordinator of the BTZ campaign) Bryan Simpson describes how,

Figure 12.1 McDonald's workers on strike, International Workers' Day, 1 May 2018

Note: Workers at five McDonald's stores went on strike to demand £10 an hour, an end to youth rates, guaranteed hours and for their right to a union to be respected.

Source: Guy Smallman.

> In 2017, following two years of direct action from Better than Zero and slow-but-steady organising from Unite hospitality members on the inside, the G1 Group [Scotland's largest hospitality employer] was forced to abandon zero-hour contracts for 2895 bar/club workers, replacing them with minimum hour contracts of at least twelve hours per week. They [the workers] were also no longer required to pay for their uniforms, will get a contract before they start working and will no longer have pay docked for marginal lateness.

A second success was clocked up in the same year when the Mooboo Bubble Tea chain was targeted after it was revealed that 'trainees' were being asked to do 40 hours of unpaid work before being considered for a job. A combination of direct action, internal organising and media exposure forced the company to capitulate within four days and commit to paying all workers from day one of training.

Both Chris (Bakers' Union) and Bryan (BTZ) were influenced by the organising model used by the American SEIU. Alan describes how, during a week he spent in Missouri, 'I was hosted through the union's links with the Fight for $15, which transformed the way I thought about organising workers. I was shocked by the racial divisions and poverty, but also touched by seeing stuff in people's homes like union posters and medals that had pride of place.'

He observed the way that the SEIU 'used a "rap" or script to talk to workers, tap into their anger and offer hope ... they found out what motivated people by asking them about their circumstances and asked "how does that make you feel"'. Bryan Simpson drew his inspiration from a two-day seminar given by Jane McAlevey (author of *No Short Cuts*, 2016, see Chapter 6) that focused on how you turn a paper and passive membership into activists and action. He explained that the key idea is finding 'organic leaders ... people that could lead from inside – not individuals with the "right ideology" but workers who were trusted, good at their jobs and socially connected and who had grievances about their workplace'.

On the ground, organisers from the Bakers' Union and Unite relied on bringing workers into action rather than servicing them with promises of cheap insurance. But the two unions operated with different opportunities and constraints. The BTZ campaign and Unite, by focusing on individual employers, had been able to recruit individuals, train them and bolster union structures with new activists. Bryan Simpson explained that this approach in these sectors has increased the number of young workers joining trade unions.

> For example, in Unite we have seen an increase of 38,000 members under the age of 30 between 2015 and 2018. The BTZ campaign was also able to tap into the creativity of workers employed in bars and restaurants to subsidise their precarious jobs in the cultural industries – musicians and actors – and to use high-profile flashmob events involving, in one case, a full choir, brass band and dancers.

The Bakers' Union, with far fewer resources that the SEIU in the US or Unite in Britain, had to work with a much more transient group of workers in what Alan described as 'fluid regional structures' and mobilising support through 'community outrage'. An example of how the Bakers' Union works in wider networks is illustrated by the Sheffield Needs a Pay Rise initiative. Inspired by the Fight for $15 campaign in the US and triggered by the publication of an academic report demonstrating that Sheffield had the lowest pay of all major cities in Britain, Sheffield Trades Council took the decision to make contact with and try to organise precarious workers. A paid organiser was hired with 50 per cent of the salary funded by the Bakers' Union and the other half by contributions from other trade unions, Labour Party wards, local community groups and charities. The brief for the organiser was to recruit to the most appropriate union, which was a departure from the 'turf war' tendency of some trade unions competing for members from the same pool. The turning point and breakthrough in the campaign was a new organiser appointed in 2020, a young woman described as a 'breath of fresh air' (president of the Trades Council) and someone who could 'kick arse' (Chris, Bakers' Union). She enlisted 20 volunteers and used 'summer patrols'[3] to go into workplaces where she succeeded in recruiting 79 per cent of the workers in three pubs into trade union membership. In addition, she recruited to or reinvigorated 'dead membership' workers from other unions. Therefore, the campaign mobilised mainly, but not exclusively, young people, including students and some 'kids off the estate', in a more open model of trade unionism – geographically based and horizontally networked.

The Covid-19 pandemic in spring 2020 was a tsunami for the hospitality industry. As restaurants, bars and clubs closed hundreds of thousands of workers were not protected under the government's furlough scheme because of their insecure contracts. Chris recalls desperate phonecalls from migrant workers in small and medium-sized firms who were in dire circumstances and not able to claim any benefits. Yet even under these circumstances unions were able to score victories. Tim Martin, the wealthy owner of JD Wetherspoon pubs, sent more than 40,000 staff a video of himself telling them that their wages had been stopped until the government's compensation scheme kicked in. The media storm created by the Bakers' Union and 95 MPs' signatures they collected forced a U-turn the day after when he agreed to pay staff. One outcome for the Bakers' Union was the creation of an 800-strong WhatsApp group. Chris said: 'I usually shy away from organising on social media, but these are difficult times ... this group might be the basis for mobilising workers in the future.' A campaign by Sheffield Needs a Pay Rise forced Papa John's pizza

Figure 12.2 Protests by workers from the Bakers' Union, supported by Sheffield Needs a Pay Rise during pandemic in July 2020, win two months' back pay

Source: Thomas Husbands.

chain in the city to pay workers the two months back pay to which they were entitled (see Figure 12.2).

Bryan explained how Unite in Scotland had a spate of wins on 'three fronts' in the first few weeks of the lockdown. They forced the reinstatement of workers in at least ten hotels or clubs, including 2,000 workers employed by the G1 Group. Chardon Hotels (owners of Holiday Inn and Holiday Inn Express) were forced to reinstate 300 workers at hotels in four Scottish cities. In other cases, employers were forced to top up wages from the 80 per cent guaranteed by the government to 100 per cent: this included the Caesars and Grosvenor Casinos. In addition, Unite won premium rates of between 120 and 165 per cent of their wages for Premier Inn staff who volunteered to work in hotels accommodating NHS workers. Previous organisational successes for Unite, the BTZ and the Bakers' Union meant that employers did not have carte blanche to lay off workers – even those who were deemed to be precarious.

Couriers: The Oldest Job in the World

Moving goods from place to place is one of the oldest jobs in the world. In the US the Pony Express, founded in 1860 to take goods and messages from the east to west coast (of the US), ran an advertisement for employees that read: 'Wanted: Young, skinny, wiry fellows not over eighteen. Must be expert riders, willing to risk death daily. Orphans preferred' (Ruttum, 2010). This is not very different from the requirements for bicycle couriers today for whom the following ad could be written: 'Wanted: super-fit people, must be expert riders, willing to risk death daily, those without personal commitments preferred.' Couriers in the so-called gig economy symbolise the fight against low pay and in particular the lack of legal protections in a sector where bogus and imposed self-employment is rife. Lesley, who had been a bike-riding courier in London since 2011 for the largest company, CitySprint, talked about how she had signed a 'heavily lawyered' contract committing her to self-employment written in language that was hard to understand.

Lesley explained that the job attracted 'waifs and strays' – people who had few opportunities in the labour market and needed quick money. However, despite this precarious work and a high turnover of couriers, the IWGB and its activists have been remarkably successful in mounting legal challenges aimed at getting employee status for this group, and have pressurised large and powerful companies into improving pay and conditions. The genesis of the Courier and Logistics branch of the IWGB was when Lesley and three other workers, angry about the number of deaths and injuries where they had no insurance and no sick pay and wages that had been stagnant for 15 years, approached the IWGB in February 2015 for help with getting organised. The 50 bicycle couriers that turned up to the first meeting testified to the accumulated grievances of workers in this sector. This sparked a campaign, initially for higher wages, focused on the market leader CitySprint which delivers to a number of major corporations and companies (for example, Goldman Sachs, Santander, *The Guardian* newspaper and the BBC). Lesley argued that 'Companies who carry the ethical badge of living wage accreditation should make sure that the couriers who do their deliveries are fairly paid', and hence one of their tactics was mounting lively flashmob indoor occupations of high-profile clients such as Google and Linklaters. Lesley commented: 'It felt powerful to be in ... expensive marble-clad lobbies, when usually we deliver through the seedy back entrance.'

On January 2016 the London Central Employment Tribunal ruled in favour of bicycle courier Maggie Dewhurst, who, supported by the IWGB, had taken CitySprint UK Ltd to court in November 2015 over her employment status. The employment judge described the self-employment status at the centre of CitySprint's business model as 'contorted', 'indecipherable' and 'window-dressing' (Butler and Osborne, 2017). This ruling opened the door for a series of other cases, against eCourier (a subsidiary of Royal Mail), Excel and Addison Lee.

The victories of bicycle couriers have been the impetus for building a Couriers and Logistics branch of the IWGB. Van drivers for eCourier, for example, face similar problems of bogus self-employment and have to pay up-front costs to the firm before they can start earning money – including £140 for van hire, £42 insurance, £5 breakdown cover and £100 for fuel. This has given confidence to other workers in the sector. In 2019, for example, cyclists as well as motorcycle rider and van driver members of IWGB working for The Doctors Laboratory, which services the routine pathology of 50 hospitals in and around London, won major concessions including formal employment status, holidays and the national minimum wage (IWGB, 2019).

Mainstreaming Sex Worker Organisation?

As early as 1980 the English Collective of Prostitutes[4] wrote to Len Murray, the then general secretary of the TUC, asking to affiliate as a union (Watson, 2019). After this request was refused, sex workers continued to be excluded from the labour movement until 2002, when the GMB established an Adult Entertainment branch that recruited across the industry, including among women directly involved in selling sex, pornography, dancing and adult chat lines. This initiative by the GMB was part of their eagerness to seek recruits from the pool of new groups of precarious workers, such as casino employees, private hire drivers, tattooists and chiropractors. However, according to Webber and Lopez (2013) the GMB quickly lost interest in their sex worker members after the imposition of austerity in the wake of the 2008 crisis led to them reprioritising their established core groups of workers. They argue that the GMB had little idea of how to relate to non-standard workers and that sex work members expressed 'dissatisfaction with the union's ability to take on board many of their issues and practices and ways of communicating', with one activist complaining that they want to 'control sex workers, rather than support them. They just take our money' (ibid.). This disillusionment was consolidated when the London region of the GMB refused to condemn the raids on and persecution of sex workers in East London in the run-up to the 2012 Olympics.

In contrast, x:talk is a radical social movement, grassroots in origin and feminist and anti-racist in politics, formed in 2006 in London to promote sex workers' rights activism and organising. Interviewee Ava Caradonna from x:talk argued that even though the GMB paraded two recognition agreements that they had struck in clubs in London and Bristol, this masked a very low membership. She went on to say that the inclusion of club owners, and even clients, in the branch made a mockery of women being able to mount any sort of independent action for better working conditions. The main political aim of x:talk is to fight for the decriminalisation of sex work through their programme Decrim Now: Campaign for Sex Workers' Rights. Laura Watson, from the English Collective of Prostitutes, argues that the starting point of the campaign for decriminalisation is:

> [N]ot asking what people think of sex work or whether sex work is uniquely degrading ... It is focused on the reality that over 70,000 women will today or tomorrow or sometime soon be exchanging sex for money and the laws governing that exchange are forcing us to do it in ways that are more isolating, dangerous and exploitative. (Watson, 2019)

A second front of activism is organising for labour rights at work, particularly in strip clubs where there is a clear employer. In 2018 x:talk joined UVW to improve conditions in clubs by trying to establish 'worker' status. This would enable those working in clubs claiming basic rights at work, such as annual leave, sick pay, a guaranteed basic wage and the right to organise and be represented by a trade union. Strippers are misclassified as self-employed, independent contractors but must follow strict rules and are forced to pay high 'house fees' to even work in clubs. Ava described successes at employment tribunals where x:talk and the UVW have won compensation for members sacked for trying to set up a trade union. In 2018 they had a large and noisy contingent on the International Women's Day strike on 8 March that shut down Soho and marched to join workers at the Picturehouse striking over pay and conditions and union recognition.[5]

Until 2017 most mainstream unions had supported the Nordic model of sex work that promotes an abolitionist position by criminalising the punters – those that purchase sex. However, mounting evidence by non-governmental organisations, for example Amnesty International, support the longstanding argument of the English Collective of Prostitutes, x:talk and other sex worker support groups that pushing sex work out of sight does not make it go away – it exposes women to danger and violence. From 2017 a sea change in the position of mainstream unions was heralded by a motion put by the Associated Society of Locomotive Engineers and Firemen to TUC conference arguing for the decriminalisation of sex work and provision of legal recognition of sex workers as workers. The motion was opposed by the TUC and other unions such as Unite and the NEU, for example, on the grounds that decriminalisation embeds exploitation and makes women more vulnerable. The notion of sex as work remains a polarised debate in the feminist movement, with divisions playing out in trade union debates. For example, a coalition of feminists (including the local branch of the Women's Equality Party) tried to get a strip club in Sheffield closed by hiring investigators to go under cover and make recordings of women working there and allegedly breaking the no-touching rule (Wolfe-Robinson, 2019b). However, despite this initial setback there has been growing support in trade unions for decriminalisation. This was reflected in an unofficial fringe meeting at Labour Party conference, A World Transformed, in 2019, with speakers from the UCU, GMB, Communication Workers' Union and UVW calling for decriminalisation. The Royal College of Nursing added their voice to this call at their 2019 conference on the grounds that it would help women access the health services they needed (RCN, 2019).

Sex work can be connected with past experiences of abuse and problems with addiction. However, the line between mainstream work and sex work is often blurred, as women take up jobs in lap-dancing clubs or on the street on a short-term basis. A parliamentary select committee received extensive evidence about how years of austerity has caused what is termed 'survival sex', specifically linked to delays in Universal Credit and draconian cuts in disability payments (Parliamentary Select Committee, 2019; Ryan, 2019). Other research has shown that the involvement of women in the industry is often a 'transitional' strategy for getting through college or university or funding expensive training (Sanders and Hardy, 2015a). Acknowledging this link between rising fees in further and higher education and sex work, the UCU overwhelmingly voted for the right of sex workers to organise and the end of criminalisation at their 2017 conference (Sanders and Hardy, 2015b; UCU, 2019b). In 2019 the GMB revived their efforts to recruit sex workers by setting up a new branch in Scotland. According to the GMB, a flagship recognition agreement for dancers and staff at a Glasgow lap-dancing venue, Seventh Heaven, has opened up collective bargaining over pay and terms and conditions (Fraser and Weir, 2019; GMB, 2019). In the light of what was the debacle of the GMB's first wave of organising, it is yet to be seen whether lessons have been learned and whether this new initiative can build a representative and democratic branch.

CONCLUSION

The games industry illustrates that exploitation lies at the heart of every employment relationship – even in 'cool' jobs. Strikes in the industry are not impossible. At Eugen Systems in France, when workers claimed that they were paid less than the amount their contract stipulated, 21 of the studio's 44 workers went on strike for six weeks (Valentine, 2018). In Britain the story of trade unions in the games development industry, thus far, is not one of a high-profile dispute, but their embryonic organisation underlines the desire to collectivise to combat poor work practices. A feature of struggle in the twenty-first century is that there are now no 'no-go' areas for trade unions. Precarious workers, such as those in hospitality and delivery services, viewed as on the periphery of the labour market and outside the scope of trade unions, have made impressive gains in fighting bogus employment and bettering their terms and conditions. The importance of these struggles, often under the radar of the media, is not the numbers involved but that the demonstration effect of workers organising under difficult conditions is an inspiration to others.

13

Capitalism's Gravediggers

In the *Communist Manifesto* Marx and Engels talk about workers as the 'gravediggers' of capitalism. Environmental catastrophe, the Covid-19 pandemic and its cataclysmic economic fallout – crises for which ordinary people are expected to pay the price – mean that debates about the combativity of the working class have never been more important. But calling the working class capitalism's gravediggers is not a description of the current state of affairs – the *actuality* of balance of power between capital and labour in the here and now – it refers, rather, to the historical *potential* of workers to bring about fundamental political change from below. Writing at a low ebb of workers' struggle this book gives a glimpse of how, even under the most challenging circumstances, workers can fight and win. It is the self-activity of workers that is key. As the sole producers of wealth, the working class has the power to break the chains of capitalism where they are forged: in the workplace. The disputes described in this book show how, during active struggle, workers experience solidarity, their confidence is transformed and they become more open to bigger ideas and the possibility of a different way of organising society.

INHERENTLY UNSTABLE, ALWAYS PRECARIOUS

Capitalism is a pinprick on the timeline of human existence, yet its pace of change and impact on the planet dramatically outstrip the economic and social change that preceded it. Marx captured the incredible technological dynamism of capitalism when he marvelled that, 'The bourgeoisie, during its scarce one hundred years, has created more massive and more colossal productive forces than have all preceding generations together' (Marx and Engels, 2015: 9). The 'whip of competition' between rival capitals drives the constant destruction of productive capacity in some places and creates new sites elsewhere, while technological change renders some types of work redundant at the same time as giving rise to new occupations. Beyond this constant kaleidoscope of work produced by incremental change, since its inception capitalism has been subject to huge convulsions, where massive expansion is followed by plunges into crisis and depression. The origins, trigger and development of each crisis of capitalism is unique and does not follow some predetermined pattern.

Structural change, with the decline of traditional bastions of trade unionism, is one reason cited for explaining why the strike rate in 2020 was the lowest since official records began. British capitalism has changed substantially in the post-war period – a process that has accelerated since the early 1970s. But the story of 'everything has changed' is fuelled by pointing to the decline of manufacturing and drawing an artificial distinction between the production of physical goods (material) and services (immaterial). The conclusion drawn is that workers engaged in these two discrete forms of production are somehow different and there is a wedge between them. This false dichotomy misses the point that the creation of surplus value does not lie at the end point of manufacturing but in modern capitalism is dispersed across a wide range of activities that blurs the boundaries between the production of goods and services. For example, physically moving goods by road has been fused with new technologies that try to reduce time in the distribution process, rebranded as logistics. But this innovation of capitalism is also a weakness. One group of workers can quickly bring sophisticated supply chains to a complete halt – car factories would lie idle and supermarkets shelves would be empty within days.

The changes in technology during my lifetime have been breathtaking. Digital technology pervades all work, from its mundane use in supermarkets to capture the spending habits of consumers and ensure 'just-in-time' deliveries to reducing the time needed for the high-technology discovery and trialling of drugs. Technology has produced new and profitable opportunities for consumption – the electronic games industry for example – and generates new forms of work. But despite dramatic prognostications that robots will cause large numbers of workers to become redundant, an argument that has been echoed throughout the history of capitalism in different forms, the evidence for 'this time it is different' is not convincing.

There has been much recent attention paid to precarious work. Using an employment legislation approach, ZHCs and agency work have been highlighted as novelties in the new landscape of employment. But since the economic crisis of 2008 a different incarnation of insecurity has intensified with workers experiencing stagnating wages and longer hours at work in order to maintain a basic standard of living, with a dramatic increase in in-work poverty after vicious reforms to benefits. The yardstick used to measure developments in the new millennium is the post-war 'standard contract'. This plays fast and loose with history. Inasmuch as it applied to workers in a subset of industries in the three decades after 1945, it never applied to all workers; building workers, dockers and workers in the film industry, for example, were notable exceptions. Claims of secure work in a previous era are blind to the fact that women and migrant workers did not reap these benefits and certain groups such as cleaners are virtually absent from this overly benign view of work stability and security.

The view of a linear and inevitable progression from stable work to insecurity misses a basic dilemma of capitalism, which is how to cut costs and increase surplus value at the same time as retaining the skilled and trained workers necessary for competition. The lack of a blueprint for managers and those who operate on behalf of capital generates the constant potential for microcosms of struggle in workplaces on multiple fronts: on wages and pensions, the intensity of work, pushing the frontiers of control, bullying and despotic managers, inequality and sexual harassment. How these battles develop is within the general frame of exploitation, but they are specific to the sector and workplace and each of them is unique.

Therefore the vicissitudes and anarchistic nature of capitalism constantly dislocates workers in terms of the jobs available, where they work and the nature of work itself, driving massive migrations within and across continents. There has been a tendency to bemoan the changes of the last five decades with an air of resignation that things are not only different in the new millennium, but that it is more difficult for workers to get organised and struggle against the rapaciousness and vagaries of neoliberal capitalism. There is a tendency to view the past through rose-tinted spectacles, with high points of industrial struggle such as those of the miners, dockers and shipbuilders in the 1970s lauded and applauded as an era of labour insurgency that will not be seen again in Britain. However, struggles in the twenty-first century show that workers organising and fighting back is still alive and kicking.

STRUGGLE IN THE NEW MILLENNIUM

Strikes are an important measure of struggle, but they are incomplete. They omit stories of workers winning ballots and employers quickly making concessions. The reductionism in reading off the level of struggle from these numbers masks the significance of disputes. The two-day strike of (mainly) women council workers in Glasgow resulted in a momentous victory on equal pay. The statistics tell us nothing about the duration of individual strikes and their resilience. In the case of the 200 women care workers in Birmingham, their fight was protracted over 18 months with 83 days of strike action before the council withdrew proposals that would have slashed their hours and wages. The all-out strike of a small handful of cleaners at BEIS would have appeared as a drop in the ocean in the strike statistics. Yet the impact of their win in terms of better terms and conditions spread to a wider group of workers and had a hugely important demonstration effect. The struggles of teachers against fully opening schools during the Covid-19 pandemic would not appear in the numbers at all as they did not go on strike. It would be foolish to overgeneralise from the handful of case study disputes between 2015 and 2020 in this book and it was never the intention to provide an epochal approach to struggle. Nevertheless, I make several key arguments.

Being a precarious worker is not a barrier to being organised and winning disputes as some have argued. For the women of Birmingham and Glasgow on permanent contracts, the source of their insecurity lay in low wages that left many in poverty. Similarly, the cleaners in London scraped a living on minimum pay but won the living wage and, at SOAS, were brought back in-house. These were migrant workers who may not have spoken English, yet who courageously and tenaciously carried out a campaign over ten years. The Bakers' Union and the broad-based intra-union campaign BTZ in Scotland have had some remarkable wins for workers who were on ZHCs and subjected to exploitative practices in restaurants and clubs.

These victories have been won both by established trade unions, in some cases the behemoths of Unison and the GMB, and by newly formed breakaway unions such as the IWGB and UVW, with only a few hundred members, who have punched above their weight. The enthusiasm and tenacity of these new labour movement organisations and the list of wins they have clocked up without having the resources, deep coffers and hundreds of thousands of members of established trade unions is hugely impressive. The anger and frustration at having the rug pulled from beneath you in a dispute by the union bureaucracy is understandable, but breaking away from established trade unions has its own problems. Being cut off from organisations with hundreds of thousands of members and a network of activists isolates some of the most militant workers. One conundrum is why Unison allegedly dampened the fight and tried to outmanoeuvre and undermine activists in one dispute and yet threw their weight behind other struggles. The answer is twofold. Trade unions walk a tightrope between trying to maintain control of disputes and preserve their machinery but at the same time need to deliver demonstrable wins to their membership to keep them on board. In the case of Glasgow, rank-and-file members demanded that their branch call action on equal pay, and, to repeat the words of the union organiser, chanted 'strike, strike, strike' at a meeting of Unison. There are limits to how far trade union bureaucracies can keep their members in check and there is room for activists to manoeuvre in the official union machine, through informal contacts with other activists and the autonomy that goes with being elected into particular positions in the workplace.

Legal barriers to industrial action in Britain, the most restrictive in Europe, have been blamed by some for the low level of struggle. Yet ballots have been won – in some cases with huge participation and majorities – even under difficult conditions. In Unite workers have won ballots and employers have quickly made concessions. University workers, council employees and cleaners have won ballots and gone on strike. This dispels pessimistic views that votes for action cannot be won in the face of restrictive Tory legislation. But it also underlines the importance of organisation from below in winning the vote for action and building solidarity within the dispute and from other workers to maintain the momentum.

The hard-won gains of women workers fighting for the equality that they were promised five decades ago are fragile. Women have borne the brunt of the 2008 financial crisis in all aspects of their employment: underemployment, cuts in the public sector and over-representation among workers on ZHCs. The Covid-19 pandemic reduced affordable childcare further and the double burden of paid work and carrying the lion's share of the domestic burden jeopardises the place of women in the labour market. Yet women have been at the forefront of most of the struggles covered in this book. This includes not only the women council workers in Glasgow and Birmingham, but also teachers, the majority of whom are young women, who fought for health and safety in their schools and communities. Women university workers, disproportionately on casualised contracts, have been at the forefront of the epic struggle against changes to pensions, casualisation and equality. The #metoo movement has exposed the continuing misogyny, sexual harassment and discrimination that women continue to face at work in the new millennium. As we have seen, despite the hip and cool image of working in the games industry, the underrepresentation of women reflects a gender politics from the dark ages.

Britain has been built by successive waves of migrant workers. They have been used to fill labour shortages, often on the bottom rung of the labour market. Yet the cleaners in London, along with sporadic strikes elsewhere, have carried on a tradition of newcomers to the labour market not being prepared to accept low pay and poor working conditions. Further, pseudo-psychological notions of a neoliberal self – whereby individualistic workers reject the collective nature of trade unions – is simply not the case. In new industries such as games development as well as old ones such as couriers – those thought to be on the margins of the labour market or in 'cool' jobs – have shown a strong desire to come together and organise against their exploitation. Young people, far from subscribing to neoliberal capitalism, have been at the forefront of organising in the hospitality sector and in schools and universities.

THE COVID-19 PANDEMIC

The cataclysmic effects of the Covid-19 pandemic on the global economy are only just beginning to unfold. Because the Conservative government's handling of the pandemic was nothing less than a complete debacle. Despite the impressive rollout of vaccinations, for which the NHS and not the government should take credit, the impact on the British economy has been disproportionately greater than that of other countries. Coupled with the decimation of public health during the years of austerity, by late January 2021 deaths from Covid-19 exceeded 100,000, the highest in Europe and the fifth highest in the world. The privatisation of the NHS supply chain and its cannibalisation by different companies who were simply handed contracts during the pandemic meant that front-line staff in the health service did not receive the PPE they desperately needed to keep them safe. The blundering incompetence of the government was part of their politics – the desire not to compromise profits – but also went beyond it. There has already been a haemorrhaging of jobs in the retail, hospitality and travel sectors. Despite the government's furlough scheme, loans and grants, there have been redundancies across the economy.

In the first few months of the pandemic, as most trade union leaders were caught like rabbits in the headlights and/or got sucked into the national rhetoric of 'we're all in it together', it was left to rank-and-file workers to fight for health and safety. Before lockdown, activists in workplaces insisted that vulnerable workers should work from home. There were walkout and refusals to work in a rash of actions over health and safety by postal workers from Aberdeen to Cornwall. Employers went on the offensive using the pandemic as an opportunity to attack wages and jobs, sometimes trying to sack workers and re-engage them on worse conditions. Despite the difficulties of organising under a lockdown, workers have not been quiescent. After Unite members launched a series of strikes at the Rolls Royce factory in Barnoldswick, Lancashire, in January 2021, management backed down from the 350 threatened compulsory redundancies and gave an assurance that production on the site was safe. In January/February 2021 British Gas engineers and call centre workers in the GMB took nine days of strike action over plans to 'fire' and 'rehire' them on much worse conditions: longer hours, reduced sick pay and less holiday. In December 2020 drug and alcohol workers employed by We Are With You in Wigan and Leigh won NHS pay rates and union recognition for Unison. This came after 26 strike days and 15 months after their first strike. They had continued to battle during lockdown. Constraints of space mean that an exhaustive account of such struggles is not possible, but these examples, alongside those discussed in earlier chapters, give a taste of some of the struggles and victories that can be read on trade union websites and in the local press, though rarely in the national media.

POLITICS AND DIALECTICS

The microdynamics of each strike is different. Seeing any of them in terms of a simple schism between the rank and file and the bureaucracy ignores their individual dynamics. There is no doubt that the strikes and campaigns discussed in this book were won from the bottom by active and militant trade union members. But organisers in trade union structures who are close to their members were important players. In the case of the Bakers' Union, Unite, Unison and the BTZ campaign, energetic and committed organisers were critical in supporting new activists. In the disputes of London cleaners, elected branch secretaries were pivotal in supporting members and maintaining action. But members had to be involved in democratic processes in a substantial and not a token way. Failure on this score explains in part the disappointing outcome of the campaign at the Sports Direct warehouse. The reliance on key full-time organisers and the community group, however dedicated, was no substitute for building a layer of activists in the workplace. The existence of a small handful of brave people prepared to stick their necks out as reps was insufficient to build a critical mass to take on a ruthless employer.

Strikes, campaigns and struggles are not set pieces between employers and unions. Neither is the relationship within unions a heroes and villains drama between conservative officialdom and radical activists from below that unfolds in a predictable way. The relationship is much more dialectical – a process of interrelationships, positive (and negative) feedback and transformation. Living struggles are not hermetically sealed bubbles. Trade union leaders and officials may initiate action from above but with a view to keeping it firmly under control within particular boundaries and ending it on their terms – usually when they feel they have flexed their muscles enough to secure a deal to sell to their members. But they are not always able to contain the workers who respond to the call for action. Strikes and industrial action are transformative. They give a sense of confidence and power to those that take part in them and contain the seeds of threatening the bureaucracy, which is at least part of the reason that trade union leaders seem to snatch defeat from the jaws of victory by making unholy deals.

One feature of the changing terrain of struggle in the new millennium is the conscious and explicit use of the 'organising model'. Activists interviewed from Unite, Unison, NEU and UCU were influenced and inspired by Jane McAlevey's book, *No Shortcuts* (2016). Sloppy organisation in terms of inaccurate membership lists is a legal disaster, and huge gaps in representation across a workplace is a poor recipe for involving members. But we need to be cautious: there is no magic bullet for mobilising workers. Recruiting new members is vital for renewing national unions and their branches, but so is involving members and union democracy. Supporting members through casework and generalising issues builds confidence in the union in terms of taking on bullying managers, but the case study disputes featured here demonstrate that it is activity and action that builds unions. The BEIS branch of cleaners began an all-out strike with a small number of members. They took a leap in the dark and recruitment accelerated when the branch was seen to be doing something. If cautious union officials or branch officers had insisted on a critical mass of strikers the dispute would not have started in the first place, let alone gone on to win a resounding victory. Categorising members in terms of their likely trade union activity, as the most mechanical version of the organising model advocates, assumes that 'organic leaders' can be known in advance. Strikes can be life changing, giving workers a sense of confidence and solidarity: in Glasgow the activists who emerged as the most militant were not necessarily the women who had come to branch meetings in the past.

A mechanical and formulaic view of workers' struggle can be dangerous and demoralising. Structural shifts in capitalism that bring about constant changes in the geography and technology of production and its crisis-ridden nature set the parameters of struggle for the working class but do not predetermine its outcome. Sometimes workers have to fight under difficult conditions and even when the odds seem to be against them. One worker said to me, 'if we organise and fight back there is no guarantee that we will win, but for sure, if we don't, we will lose – that is the only thing that is certain'. But there is a bigger picture. Wider politics, in the form of the resistance of and protests by other groups of workers at home or internationally, are important for struggle – they are inspirational and raise the bar of what can be demanded and won.

Workers on strike are open to seeing the world in a different way. Paul McGarr points out how the Tunisian and Egyptian Revolutions of the late 2010 and early 2011 spilled into the consciousness of activists: 'in the movement at the time there was much talk of "if we can topple Mubarak, we can fight Cameron" or "we should turn Trafalgar Square into Tahrir Square"' (2016: 115). It was not the case that millions of workers in Britain shared this connection, but along with the student revolt against higher fees in 2010 it changed the mood of a layer of activists and representatives. This meshes with how socialists can play a decisive role in the microcosm of the workplace. Not only are they less constrained by the narrow straitjacket of industrial relations, more committed to action from below and solidarity between groups of workers, but they are a bridge between external events and the workers they represent or work alongside by generalising resistance elsewhere.

CAPITALISM ON THE BRINK

The exploitation intrinsic to capitalism renders the workplace an arena for low-level guerrilla warfare and constant skirmishes. These have the potential to explode into larger struggles from the tinder of long-running grievances or in response to sudden attacks by employers on jobs, wages and working conditions. But beyond the drama of individual workplaces and microcosms of struggle endemic to capitalism there are big challenges common to all workers. The catastrophe of global warming and climate change is an immediate and urgent threat to the planet. Actions by governments have been grudging, piecemeal and disingenuous, driven by a desire to be seen to be doing something, but unprepared to make any changes that disrupt and threaten profits. The dire consequences of looming environmental disaster cannot be averted without a complete reorganisation of society and changed priorities.

At the time of writing the depth of the global economic crisis is unknown. But what is certain is that the fallout from the Covid-19 epidemic will set the stage for future fights as the ruling class attempts to foist the crisis onto workers who are being asked to pay in terms of their jobs and wages and health and safety. This is heaped on the austerity of the post-2008 financial crisis from which living standards and welfare services have never recovered. It could not be clearer who the essential workers are. It is the doctors, nurses and care workers, teachers, bus and train drivers, supermarket workers and refuse collectors who keep society running. Investment bankers are notable by their absence from this list. Yet essential workers are the least rewarded and struggle to do their jobs under difficult and often dangerous conditions.

The political and economic fallout of the 2008 crisis compounded by the havoc of the Covid-19 pandemic and consequent burden on working-class people can take a darker direction. The success of populist, racist and xenophobic parties in Europe and beyond is a sign of the despair of ordinary people and an ominous and salutary reminder of a different and divisive direction of travel. The alternative project of emancipating the working class involves going beyond mitigating exploitation and negotiating its terms within the parameters of trade unions: it demands the evisceration of exploitation itself through a root and branch change to society. Being part of and winning their disputes is a life-changing experience for workers not only in terms of material improvements to their quality of life, but in restoring their dignity and bolstering their confidence.

Radical trade unionism is important but insufficient. Struggles need to be linked and generalised into a much wider and fundamental economic and political challenge to the system itself. The incredible developments in technology and science – drug discovery, pushing digital frontiers and the increased sophistication of robots – have huge liberatory potential for curing debilitating illnesses and for removing the drudgery of work. But these benefits will not be realised when production is in the hands of private capital, driven by profit and not need. We saw how quickly firms producing yet another iteration of some consumer durable could quickly redirect their resources to producing ventilators, or switch to making PPE. This can only be done by workers from below, collectively and democratically deciding in their workplace how to produce things, what things to produce and for whom. The stark choice posed by Rosa Luxemburg has never been more apposite: socialism or barbarism.

Notes

CHAPTER 2

1. See Mirowski and Plehwe (2009) and Peck (2010) for a detailed discussion of the development of neoliberalism.
2. See Hardy (2016) for a critique of financialisation as a new stage of capitalism.
3. Property portfolios of care homes were sold off to an investor and then rented back at pre-agreed rent increases that were unsustainably high – sometimes all within the same company (*Accountancy Daily*, 2016).
4. Produced by a joint committee of the Department for Business, Energy and Industrial Strategy and the Department of Work and Pensions (BEIS and Work and Pensions Committees, 2018).

CHAPTER 3

1. Aglietta (1976) is often credited with popularising the term post-Fordism. See Brenner and Glick (1991) for a robust critique.
2. The term post-industrial originated with Alain Touraine in 1971 and was popularised by Daniel Bell (1973) to highlight the rise of the service sector and increased salience of knowledge and those that did 'intellectual work'.
3. The Group of Seven is an international intergovernmental economic organisation consisting of seven major developed countries: Canada, France, Germany, Italy, Japan, the UK and the US, which are the largest IMF-advanced economies in the world.
4. The *Grundrisse* is a series of seven notebooks rough-drafted by Marx during the winter of 1857–8. The manuscript was lost and then first published, in the German original, in 1953.
5. Conversation with Bob Jeffrey, president of the Sheffield Needs a Pay Rise campaign. See tinyurl.com/y6syd9tl (last accessed 8 February 2021).

CHAPTER 4

1. See Alberti et al. (2018) for a full discussion of the problems in defining precarious work.
2. Forty-seven years later the campaign's researcher Eileen Turnbull unearthed fresh evidence that persuaded the Criminal Cases Review Commission to refer the case to the appeal court (*Socialist Worker*, 2020). In 2019 50 trade unionists won £1.9 million in compensation from building firms (Evans,

2019). In March 2021, after more than four decades of campaigning, court of appeal judges overturned the convictions of the Shrewsbury building workers, including actor Ricky Tomlinson.

3. Entitlements include: a right to the national minimum wage, holiday pay, protection against discrimination and the right to be treated fairly even if working part-time.

CHAPTER 5

1. For example, *I'm All Right Jack* and *The Angry Silence* were overtly anti-union films.
2. Officially named the Royal Commission on Trade Unions and Employers' Associations it was set up by the Labour government in 1965 (reported in 1968) and known by the name of its chair, Lord Donovan (Banks, 1969).
3. Known as 'In Place of Strife' it was proposed by the secretary of state for employment and productivity, Barbara Castle. Among its proposals were plans to force unions to ballot before a strike was held and the establishment of an industrial board to enforce settlements in industrial disputes. Splits over whether union action should be restricted in the Labour government's cabinet meant that it never became legislation.
4. Drawn up by wealthy right-wing backbench Conservative MP Nicholas Ridley, a founding member of the Selsdon Group of free market Conservatives. In the report he proposed how the next Conservative government could fight, and defeat, a major strike in a nationalised industry.
5. GCHQ (Government Communication Headquarters) is the government's intelligence and security organisation.
6. Written as a co-operative venture by sacked dockers with Jimmy McGovern and Irvine Welsh overseeing the script.
7. The FBU voted to disaffiliate from the Labour Party after the hostile treatment they received from the Blair government in their 2002–3 pay dispute. They reaffiliated in 2015 after the election of Jeremy Corbyn as leader. The RMT was expelled when members voted overwhelmingly to ignore an ultimatum to stop supporting Labour's electoral rivals, including the Scottish Socialists.
8. See debate between O'Brien (2018) and Lyddon (2018) for historical detail.
9. McCluskey (2018).

CHAPTER 6

1. The NATFHE, who represented workers in what were, before 1992, polytechnics under the auspices of local authorities and the AUT, who represented teaching, research and administrative workers in the 'old' universities. These two trade unions merged to form the UCU on 1 June 2006.

2. Affiliated trade unions pay an annual fee to the Labour Party; in return they elect twelve of the 32 members of Labour's National Executive Committee and 50 per cent of delegates to the Labour Party conference.
3. There are other examples in Britain of breakaway unions such as the United Clothing Workers Union in the 1920s. See Lerner (1961).
4. TGWU became Unite the Union.

CHAPTER 7

1. Shared parental leave, introduced in April 2015, allows mothers to transfer some of their £151.20 (in 2020) statutory pay to their partners two weeks after the birth (or adoption) of their child.
2. It included the union that represented civil servants (the then National Association of Local Government Officers), the British Medical Association, the London County Council Staff Association and the Communist Party, among others.
3. Under the Factory Act women could not work night shifts in factories and this section was not repealed until 1986.
4. This can be found at tinyurl.com/20qfrh5k (last accessed 7 February 2021).

CHAPTER 8

1. Czech Republic, Estonia, Hungary, Latvia, Lithuania, Poland, Slovakia and Slovenia.
2. *Deutscher Gewerkschaftsbund* and the *Österreichischer Gewerkschaftsbund.*
3. The interviews and anecdotes in this chapter are drawn from the Cross Border Trade Union Collaboration, a project funded by the Economic and Social Research Council with Ian Fitzgerald (February 2006 to April 2009), and my research in Poland from 1993 onwards.
4. OPZZ (*Ogólnopolskie Porozumienie Związków Zawodowych*) was the official 'yellow' union of the Communist Party. There was extreme hostility between them and *Solidarność* in the early 1990s. As OPZZ has morphed into a 'proper' non-secular trade union there is some cooperation between them, particularly at the level of the workplace.
5. Very little has been written about language federations in the early twentieth century. But the arrival of migrants with little English led to the formation of branches based on country of origin and language group. See https://en.wikipedia.org/wiki/Non-English_press_of_the_Socialist_Party_of_America.
6. The Union Learning Fund was set up by the government in 1998 to promote activity by unions in support of the government's objective of creating a 'learning society'. The role of the Union Learning Representative is given recognised status similar to that of union health and safety representatives. See www.unionlearningfund.org.uk.

CHAPTER 9

1. Unite the Union, commonly known as Unite, is a British and Irish trade union formed on 1 May 2007 by the merger of Amicus and the TGWU. With just over 1.2 million members, it is the second largest trade union in the UK.
2. Mother of the chapel and father of the chapel are the titles in Britain referring to a shop steward representing members of a trade union in a printing office or in journalism. In the printing trade a chapel is the traditional name given to a meeting of compositors.

CHAPTER 10

1. 'And Did Those Feet in Ancient Time' is a poem by William Blake from the preface to his epic *Milton: A Poem in Two Books*, one of a collection of writings known as the *Prophetic Books*. See Cox (2004: 289).
2. With an estimated net worth of £2.3 billion Mike Ashley has done extremely well out of the workers at the Shirebrook warehouse. It has enabled him to go on a shopping spree of ailing high-street retailers. He has majority or minority shares in: Dunlop Slazenger, Karrimor, Kangol, Lonsdale, the Blacks Leisure Group, sofa.com, Evans Cycles and GAME Digital. In a high-profile deal in 2018 he bought House of Fraser. To finance his empire even further he sold the site at Shirebrook for £120 million and is leasing it back for fifteen years (*Independent*, 2019; Davies, 2019).
3. See also a report on the experiences of workers produced by the Derbyshire Unemployed Workers' Centre (Ball et al., 2017)

CHAPTER 11

1. These were researchers, teaching staff and academic-related and professional staff (senior administrators, computing and library workers) in the pre-92 universities who are part of the Universities Superannuation Scheme (USS). The Further and Higher Education Act 1992 made changes to the funding and administration of further education and higher education within England and Wales. The most visible result was to allow 35 polytechnics to become universities, often referred to as the 'new universities' or 'post-1992 universities'.
2. The AUT represented researchers, teaching staff and administrators in the pre-92 universities and NATFHE drew members from polytechnics (post-92 universities) and further education colleges.
3. https://uculeft.org/.
4. USS Briefs, https://ussbriefs.com/briefs/.

5. In January 2021 some supporters of this group (along with other new rank-and-file activists) launched itself as the UCU Commons.

CHAPTER 12

1. In 2017 there were 2,175 games development firms in the UK, the vast majority concentrated in London (563) with Manchester in second place (85). In addition, other cities and towns have developed clusters: Sheffield (36), Birmingham (33), Liverpool/Nottingham (32) and Edinburgh (UKIE, 2018).
2. In Britain the games development industry receives more venture capital than in any other European country. Multinational firms that have invested include: Microsoft, Activision Blizzard, Sony Interactive Entertainment, Konami, Take-Two, Warner Brothers, EA and Ubisoft.
3. The idea of summer patrols comes from the Norwegian trade union centre LO-Norway. Every summer for over thirty years LO-Norway has sent trade union activists into workplaces where young people are likely to be working in worse conditions. The TUC has sent delegates to Norway and has also used this strategy in the Yorkshire and Humber region. See https://tinyurl.com/22xustmw (last accessed 13 February 2021).
4. At the end of 1975 the new prostitutes' branch of the Wages for Housework campaign became the English Collective of Prostitutes (ECP). The ECP was formed as part of the highly politicised prostitutes' rights movement that emerged in Europe in the mid-1970s. The 1975 prostitutes' strike in France and the subsequent formation of the French Prostitute Collective inspired the formation of a similar organisation in England. See https://prostitutescollective.net/ecp-history/ (last accessed 13 February 2021).
5. Workers at four Picturehouse cinemas in London, members of the Prospect union, were on strike for a living wage, company maternity and paternity allowances, improved sick pay and union recognition.

References

Accountancy Daily (2016) 'Care homes under pressure as property leaseback costs soar'. tinyurl.com/3wvjleb2 (last accessed 6 February, 2021).

Aglietta, M. (1976) *A Theory of Capitalist Regulation: The US Experience*, London: Verso.

Alberti, G. (2016) 'Mobilizing and bargaining at the edge of informality: "The 3 Cosas campaign" by the outsourced migrant workers at the University of London', *The Journal of Labor and Society*, 19: 81–103.

Alberti, G., Bessa I., Hardy, K., Trappman, V. and Umney, C. (2018) 'In, against and beyond precarity: work in insecure times', *Work, Employment and Society*, 32(3): 447–57. Anitha, S. and Pearson, R. (2018) *Striking Women: Struggles and Strategies of South Asian Women Workers From Grunwick to Gate Gourmet*, London: Lawrence & Wishart.

Arntz, M., Gregory, T. and Zierahn, U. (2016) 'The risk of automation for jobs in OECD countries: a comparative analysis', *OECD Social, Employment and Migration Working Papers*, No. 189, Paris: OECD Publishing. tinyurl.com/jfiexiot (last accessed 6 February 2021).

Atkinson, W. and Randle, K. (2014) '"Sorry mate you're finishing tonight": a historical perspective on employment flexibility in the UK film industry', *Work Organisation, Labour and Globalisation*, 8(1): 49–68.

Ball, M., Hampton, C., Kamerade, D. and Richardson, H. (2017) *Agency Workers and Zero Hours: The Story of Hidden Exploitation*, report published by Derbyshire Unemployed Workers Centre. http://usir.salford.ac.uk/id/eprint/43586/1/DUWC%20Research%20Report%208698_p1.pdf (last accessed 14 February 2021).

Banks, R.F. (1969) 'The reform of British industrial relations: the Donovan Report and the Labour government's policy proposals', *Erudite*, 24(2): 333–82, https://id.erudit.org/iderudit/028022 (last accessed 12 February 2021).

Barker, C. (1970) *The Pilkington Strike*, Socialist Worker pamphlet, London: International Socialists.

Barnsley, T. (2010) *Mary Macarthur and the Chainmakers' Strike of 1910*, London: Bookmarks.

BBC (2012) 'Miliband shifts immigration policy, saying Labour "got it wrong"', BBC News, 22 June. www.bbc.co.uk/news/uk-politics-18539472 (last accessed 13 April 2021).

BBC (2019) 'Gaming worth more than video and music combined', BBC News, 3 January. tinyurl.com/mjd68te5 (last accessed 13 February 2021).

BECTU (2019) 'Survey of UK gaming industry reveals long hours culture and bullying claims', 9 August. tinyurl.com/4cyvxw9x (last accessed 13 February 2021).

BEIS (Business, Energy and Industrial Strategy) (2019) 'Automation and the future of work', House of Commons Business, Energy and Industrial Strategy Committee, Twenty-Third Report of Session 2017–19, HC 1093, 18 September.

BEIS (Business, Energy and Industrial Strategy) Committee (2018) 'The impact of Brexit on the pharmaceutical sector', Ninth Report of Session 2017–19, House of Commons. tinyurl.com/3slugjug (last accessed 7 February 2021).

BEIS (Business Energy and Industrial Strategy) and Work and Pensions Committees (2018) 'Carillion', House of Commons, HC 769, 16 May. tinyurl.com/49bob9a2 (last accessed 6 February 2021).

Bell, D. (1973) *The Coming of Post-industrial Society: A Venture in Social Forecasting*, New York: Basic Books.

Bergfeld, M. (2018) '"Do you believe in life after work?" The University and College Union strike in Britain', *Transfer*, 24(2): 233–6.

Beveridge, W. (1942) *Social Insurance and Allied Services: Report*, London: His Majesty's Stationery Office. http://pombo.free.fr/beveridge42.pdf (last accessed 7 February 2021).

Beynon, H. (1973) *Working for Ford*, London: Allen Lane.

Blackwell, R. (2017) *Forty Miles of Bad Road: The Stars Campaign for Interracial Friendship and the 1958 Notting Hill Riots*, London: Redwords.

Böheim, R. and Kepler, J. (2009) 'Dependent self-employment: workers between employment and self-employment in the UK', *Journal for Labour Market Research*. tinyurl.com/upuucu7s (last accessed 7 February 2021).

Booth, R. (2018) 'DPD courier who was fined for day off to see doctor dies from diabetes', *The Guardian*, 5 February. tinyurl.com/49fyywpz (last accessed 7 February 2021).

Boston, S. (2015) *Women Workers and Trade Unions*, London: Lawrence & Wishart.

Bourdieu, P. (1963) *Travail en travailleurs en Algerie*, Paris: Mouton & Co.

Bourdieu, P. (1998) 'The essence of neoliberalism', *Le Monde Diplomatique*, December, https://mondediplo.com/1998/12/08bourdieu (last accessed 7 February 2021).

Bread and Roses (2000) Film, directed by Ken Loach, FilmFour Distribution.

Brenner, R. and Glick, M. (1991) 'The regulation approach: theory and history', *New Left Review*, July–August, 1: 188.

Briggs, A. and Saville, J. (1971) *Essays in Labour History, 1886 to 1923*, London and Basingstoke: Macmillan Press.

Brown, W. (2011) 'Industrial relations in Britain under New Labour, 1997–2010: a post-mortem', *Journal of Industrial Relations*, 53(3): 402–13.

Brown, W. (2015) *Undoing the Demos: Neoliberalism's Stealth Revolution*, New York: Zone Books.

Butler, S. and Osborne, H. (2017) 'Courier wins holiday pay in key tribunal ruling on gig economy', *The Guardian*, 6 January, www.theguardian.com/business/2017/jan/06/courier-wins-holiday-pay-in-latest-key-tribunal-ruling-for-gig-economy (last accessed 9 April 2021).

CAB (Citizens Advice Bureau) (2015) *Who Are the Self Employed?* Report, 21 March 2015. tinyurl.com/5xh655cg (last accessed 7 February 2021).

Callinicos, A. (1982) 'The rank and file movement today', *International Socialism*, Autumn, 2(17), www.marxists.org/history/etol/writers/callinicos/1982/xx/rfmvmt.html (last accessed 9 April 2021).

Callinicos, A. and Simons, M. (1985) *The Great Strike: The Miners' Strike of 1984–5 and Its Lessons*, London: Socialist Worker Publications.

Cambridge UCU (2018) 'Cambridge dons dancing', https://www.youtube.com/watch?v=HyJgYtbMc1k (last accessed 13 February 2021).

Charlton, J. (1999) *It Just Went Like Tinder: The Mass Movement and New Unionism in Britain 1889*, London: Redwords.

Choonara, J. (2019) *Insecurity, Precarious Work and Labour Markets: Challenging the Orthodoxy*, London: Routledge.

CIPD (Chartered Institute of Personnel and Development (2020) 'Employment status Q&As', November. tinyurl.com/1dbckxeb (last accessed 7 February 2021).

Clark, D. (2020) 'Foodbanks: emergency food needed United Kingdom (UK) 2008–2019', Statista, 17 January. tinyurl.com/1h8doiz5 (last accessed 7 February 2021).

Clarke, S. and Cominetti, N. (2019) 'Setting the record straight: how record employment has changed the UK', Resolution Foundation Report, January. tinyurl.com/269u8 (last accessed 7 February 2021).

Cliff, T. and Gluckstein, D. (1986) *Marxism and Trade Union Struggle: The General Strike of 1926*, London and Chicago: Bookmarks.

Copsey, N. (2017) *Anti-Fascism in Britain*, 2nd edition, London: Routledge.

Corlett, A. and Finch, D. (2016) 'Double Take: Workers with multiple jobs and reforms to National Insurance', Resolution Foundation Briefing, November. tinyurl.com/5yktwy68 (last accessed 7 February 2021).

Cox, M., ed. (2004) '1808', in *The Concise Oxford Chronology of English Literature*, p. 289, Oxford, Oxford University Press.

Coyle, D. and Quah, D. (2002) *Getting the Measure of the New Economy*, Place Industrial Society, May.

Darlington, R. (2009) 'Leadership and Union Militancy: The Case of the RMT', *Capital and Class*, 33, (3): 3–32.

Darlington, R. (2014a) 'The rank and file and the trade union bureaucracy', *International Socialism*, Spring, 142: 57–82. tinyurl.com/3cmqvfhq (last accessed 7 February 2021).

Darlington, R. (2014b) 'Strike waves, union growth and the rank-and-file/bureaucracy interplay: Britain 1889–1890, 1910–1913 and 1919–1920', *Labor History*, 55(1): 1–20.

Darlington, R. and Lyddon, D. (2001) *Glorious Summer: Class Struggle in Britain 1972*, London, Chicago and Sydney: Bookmarks.

Davies, R. (2017) 'Brothers jailed for trafficking people from Poland to work at Sports Direct', *The Guardian*, 23 January. tinyurl.com/1efk8hkm (last accessed 7 February).

Davies, R. (2019) 'Game Digital agrees £52m takeover by Mike Ashley's Sports Direct', *The Guardian*, 21 June. tinyurl.com/2wmp9pct (last accessed 7 February 2021).

Davies, W. (2016) *The Happiness Industry: How the Government and Big Business Sold Us Well-being*, London and New York: Verso.

Davies, W. (2017) *The Limits of Neoliberalism: Authority, Sovereignty and the Logic of Competition*, London: SAGE.

Davis, M. (2020) 'An historical introduction to the campaign for equal pay, union history, winning equal pay: the value of women's work', TUC Library Collections, London Metropolitan University. tinyurl.com/wrejsntr (last accessed 7 February 2021).

Dealessandri, M. (2019) 'State of the union: is it time for devs to unionise', *MCV/DEVELOP Magazine: Video Games Industry Analysis and Insight*. tinyurl.com/1xcqnh2b (last accessed 13 February 2021).

Denvir, J. (1894) *The Irish in Britain*, London: Kegan-Paul reproduced by Bibliolife.

Department for Culture Media and Sport (2016) 'Creative industries estimated', 26 January. tinyurl.com/1mlje3mh (last accessed 7 February 2021).

Dockers (1990) TV drama, written by Jimmy McGovern and Irvine Welsh (and sacked dockworkers), Channel 4, available as DVD.

Doogan, K. (2009) *New Capitalism: The Transformation of Work*, Cambridge: Polity Press.

Draper, H. (1978) *Karl Marx's Theory of Revolution*, Vol. 2, New York and London: Monthly Review Press.

Drayton, E. and Waltmann, B. (2020) 'Will universities need bailout to survive the Covid-19 crisis?', Briefing Note BN300, Institute of Fiscal Studies, July. tinyurl.com/1fvrnyfy (last accessed 13 February 2021).

Dresser, M. (1986) *Black and White on the Buses: The 1963 Colour Bar Dispute*, Bristol: Pressgang Co-operative Ltd.

Engels, F. (1881) 'Trades Unions', *The Labour Standard*, 28 May. tinyurl.com/1btc229k (last accessed 7 February 2021).

EHRC (Equality and Human Rights Commission) (2010) *Inquiry into Recruitment and Employment in the Meat and Poultry Processing Sector Report*. tinyurl.com/u4bp8rsb (last accessed 7 February 2021).

Evans, R. (2019) '50 blacklisted trade unionists win £1.9 million from building firms', *The Guardian*, 14 May. tinyurl.com/1a9gv846 (last accessed 7 February 2021).

Farrell, S. and Butler, S. (2016) 'Sports Direct ditches zero-hours jobs and ups worker representation', *The Guardian*, 6 September. tinyurl.com/n7jr55rk (last accessed 17 February 2021).

Fawcett Society (2020) *BAME Women and Covid-19: Research Evidence.* tinyurl.com/1xnbpoks (last accessed 10 February 2021).
Fine, J. (2006) *Worker Centers: Organizing Communities at the Edge of a Dream*, Ithaca, NY: Cornell University.
Fitzgerald, I. and Hardy, J. (2010) '"Thinking outside the box"? Trade union organizing strategies and Polish migrant workers in the United Kingdom', *British Journal of Industrial Relations*, 48(1): 131–50.
Ford, M. (2016) *The Rise of the Robots: Technology and the Threat of Mass Unemployment*, London: Oneworld Publishing.
Foucault, M. (2008) *The Birth of Biopolitics: Lectures at the College de France, 1978–79*, trans. Graham Burchell, New York: Palgrave Macmillan.
Francis-Devine, B. and Pyper, D. (2020) 'The gender pay gap', Briefing Paper 7068, 20 November, London: House of Commons Library. tinyurl.com/r7i2rjsf (last accessed 7 February 2021).
Fraser, G. and Weir, C Z. (2019) 'The Glasgow lap dancers who joined a union', BBC News, 18 July. tinyurl.com/yfb6lgbh (last accessed 13 February 2021).
Frey, C. B. and Osborne, M. A. (2017) 'The future of employment: how susceptible are jobs to computerisation?', *Technological Forecasting and Social Change*, 114: 254–80. tinyurl.com/3kyl2cxh (last accessed 7 February 2021).
Frow, R. and Frow, E. (1989) *Political Women, 1800–1850*, London: Pluto Press.
Fryer, P. (1984) *Staying Power: The History of Black People in Britain*, London: Pluto Press.
Gill, T. (2013) 'The Indian Workers' Association Coventry 1938–1990: political and social action', *South Asian History and Culture*, 4(4): 554–73.
GMB (2017) 'Insecure: tackling precarious work and the gig economy', Central Executive Council Special Report. tinyurl.com/t2c9bhhg (last accessed 7 February 2021).
GMB (2019) 'Ground breaking deal struck with lap dancing club', 21 June. tinyurl.com/4sxevugf (last accessed 13 February 2021).
Goldthorpe, J. H., Lockwood, D., Bechhofer, F. and Platt, J. (1969) *The Affluent Worker in the Class Structure*, Cambridge: Cambridge University Press.
Goodley, S. (2017) 'Sports Direct agency Transline preparing for insolvency', *The Guardian*, 26 April. tinyurl.com/4zxjky42 (last accessed 7 February 2021).
Gorz, A. (1997) *Farewell to the Working Class*, London: Pluto Press.
Grady, J. and Simms, M. (2019) 'Trade unions and the challenge of fostering solidarities in an era of financialisation', *Economic and Industrial Democracy*, 40(3): 490–510.
Griffiths, A. and Wall, S. (2004) *Applied Economics*, 10th edition, London: Prentice Hall.
Groves, S. and Merritt, V. (2018) *Trico: A Victory to Remember*, Chadwell Heath: Lawrence & Wishart.

Guardian, The (2016) '*The Guardian* view on the trade union bill: unprincipled and unnecessary – editorial', 2 May.

Hallas, D. (1980) 'Trade unions and revolution: a response to Richard Hyman', *International Socialism*, 2(8): 80–4. tinyurl.com/y3ry5ptq (last accessed 7 February 2021).

Hammer, N. (2015) 'New industry on a skewed playing field: supply chain relations and working conditions in UK garment manufacturing; focus area – Leicester and the East Midlands', *Centre for Sustainable Work and Employment Futures,* University of Leicester. tinyurl.com/2uvxyulg (last accessed 7 February 2021).

Harari, D. (2020) 'Productivity: key economic indicators', Research Briefing 02791, House of Commons, 8 April 2010. tinyurl.com/yz8fbue7 (last accessed 7 February 2021).

Hardt, M. and Negri, A. (2000) *Empire*, Cambridge, MA and London: Harvard University Press.

Hardy, J. (2013) 'New divisions of labour in the global economy', *International Socialism*, Winter, 137: 101–26.

Hardy, J. (2016) 'Radical economics, Marxist economics and Marx's economics', *International Socialism*, 149: 71–100.

Hardy, J. and Choonara, J. (2014) 'Neoliberalism and the British working class: a reply to Neil Davidson', *International Socialism*, 140: 103–30.

Hardy, J., Shelley, S., Calveley, M. and Kubisa, J. (2016) 'Scaling the mobility of health workers in an enlarged Europe: an open political-economy perspective', *European Urban and Regional Studies*, 23(4): 798–815.

Harvey, D. (2005) *A Brief History of Neoliberalism*, Oxford: Oxford University Press.

Hauge, J. and O'Sullivan, E. (2019) 'Inside the black box of manufacturing: conceptualising and counting manufacturing in the economy', report prepared for Department of Business, Energy and Industrial Strategy, Department of Engineering, Cambridge: University of Cambridge.

Health and Safety Executive (HSE) (2020) 'Work related stress, anxiety or depression statistics in Great Britain, 2019'. tinyurl.com/y5udsjjt (last accessed 7 February 2021).

Hearn, J. and Bergos, M. (2010) 'Learning from the cleaners? Trade union activism among low paid Latin American migrant workers at the University of London', Working Paper No. 7, Identity, Citizen and Migration Centre, University of Nottingham.

Heath, N. (2014) 'A history of the IWW in Britain (1913–1920)', 25 June. https://iww.org.uk/news/902/ (last accessed 7 February 2021).

Hinton, J. and Hyman, R. (1975) *Trade Unions and Revolution: The Industrial History of the Early British Communist Party*, London: Pluto Press.

Hobbs, M. (1973) *May Hobbs: Born to Struggle*, London: Quartet Books.

Home Affairs Select Committee (2016) 'Proceeds of crime', Fifth Report of Session 2016–17, House of Commons, 29 June 2016. tinyurl.com/2uno2rzc (last accessed 7 February 2021).

Home Office (2007) 'The economic and fiscal impacts of immigration', cross departmental submission to the House of Lords Select Committee on Economic Affairs, October. tinyurl.com/1pqpb9ms (last accessed 7 February 2021).

House of Commons Select Committee (2016) 'Employment practices at Sports Direct', Third Report of Session 2016–17, HC 219, London: House of Commons, 22 July. tinyurl.com/yj2wxqqq (last accessed 7 February 2021).

House of Lords (2008) 'The economic impact of immigration', Economic Affairs: First Report, Select Committee on Economic Affairs. tinyurl.com/1rakpu2l (last accessed 7 February 2021).

Hutton, W. (2004) 'Don't weep for our lost factories', *The Observer*, 19 December.

Hutton, G. and Shalchi, A. (2021) 'Financial services: contribution to the UK economy', House of Commons Library, Briefing Paper 6193, 31 July. https://commonslibrary.parliament.uk/research-briefings/sn06193/ (last accessed 3 April 2021).

Hyman, R. (1980) 'British trade unionism: post-war trends and prospects', *International Socialism*, Spring, 2(8): 64–79. https://marxists.architexturez.net/history/etol/newspape/isj2/1980/no2-008/hyman.html (last accessed 13 February 2021).

Independent, The (2019) 'Mike Ashley: what does Sports Direct CEO earn and how much is he worth?', 26 March. tinyurl.com/yo8p3kv3 (last accessed on 7 February 2021).

I'm All Right Jack (1959) Film, directed by John Boulting, Charter Film Productions, United Kingdom.

ISS (2019) 'Update on partnership with Novartis'. www.issworld.com/press/news/2019/05/27/update-on-partnership-with-novartis (last accessed 7 February 2021).

Johnson, B. (2015) *Who Dips in the Tin? The Butty System in the Nottinghamshire Coalfield*, Nottingham and Derbyshire Labour History Society, Occasional Pamphlet 2, March.

Katsomitros, A. (2018) 'The emerging university bonds market', *World Finance*, 20 April. tinyurl.com/2dz5tty9 (last accessed 7 February 2021).

Labour Focus on Exploitation (2021) 'FLEX's participatory research highlights systemic labour abuses experienced by UK cleaners'. www.labourexploitation.org/news/new-report-flex%E2%80%99s-participatory-research-highlights-systemic-labour-abuses-experienced-uk (last accessed 12 April 2021).

Lane, T. and Roberts, K. (1971) *Strike at Pilkington's*, London: Collins/Fontana.

Lavalette, M. and Kennedy, J. (1996) *Solidarity on the Waterfront: The Liverpool Lockout of 1995/96*, Birkenhead: Liver Press.

Lazzarato, M. (1996) 'Immaterial labor', in P. Virno and M. Hardt (eds), *Radical Thought in Italy: A Potential Politics*, Minneapolis: University of Minnesota Press, 133–47.

Leeds – United! (1974) Play for Today, BBC, directed by Roy Battersby and written by Colin Welland, first aired 31 October.

Lenin, V. I. (1970 [1921]) *'Left Wing' Communism an Infantile Disorder*, originally published in the People's Republic of China, 1965. Reprinted in the US by Red Star Publishers, 2016.

Lerner, S. W. (1961) *Breakaway Unions and the Small Trade Union*, London: George Allen & Unwin.

Lewenhak, S. (1977) *Women and Trade Unions: An Outline of Women in the British Trade Union Movement*, London and Tonbridge: Ernest Benn Limited.

Luxemburg, R. (1906) *The Mass Strike, the Political Party and the Trade Unions*, Marxist Educational Society of Detroit, 1925, reproduced online by the Rosa Luxemburg Internet Archive (marxists.org) 1999. tinyurl.com/2ua9r85c (last accessed 7 February 2021) .

Lyddon, D. (2015a) 'The changing pattern of UK strikes, 1964–2014', *Employee Relations*, 37(6): 733–45.

Lyddon, D. (2015b) 'Striking facts about the "Winter of Discontent"', *Historical Studies in Industrial Relations*, 36: 205–18.

Lyddon, D. (2018) 'Why trade union legislation and the Labour Party are not responsible for the decline in strike activity', *International Socialism*, Spring, 158: 191–215.

Lyddon, D. (2020) TUC history online. tinyurl.com/5cpdlnq6 (last accessed 7 February 2021).

Mahamdallie, H. (2007) 'Muslim working class struggles', *International Socialism*, Winter, 113. www.isj.org.uk/?id=288.

Mankelow, R. and Wilkinson, F. (1998) 'Industrial relations in iron and steel, shipbuilding and the docks 1930–1960', in Whiteside, N. and Salais, R. (eds), *Governance, Industry and Labour Markets in Britain and France: The Modernising State in the Mid-twentieth Century*, London: Routledge.

Marx, K. (1977 [1887, first published 1954]) *Capital: A Critique of Political Economy*, Vol. 1, London: Lawrence and Wishart.

Marx, K. (1973 [first published translation by Pelican Books reprinted 1993]) *Grundrisse*, London: Penguin Classics.

Marx, K. and Engels, F. (2015 [1888]) *The Communist Manifesto*, London, Penguin Classics.

Mason, P. (2015) *Postcapitalism: A Guide to Our Future*, London: Penguin Books.

McAlevey, J. F. (2016) *No Shortcuts: Organising for Power*, Oxford and New York: Oxford University Press.

McCluskey, L. (2018) Speech to Unite Policy Conference, 5 July. www.youtube.com/watch?v=Rop7H7162W4 (last accessed 5 January 2020).

McDowell, L. (2013) *Working Lives: Gender, Migration and Employment in Britain, 1945 to 2007*, Chichester: Wiley-Blackwell.

McDowell, L. (2016) *Migrant Women's Voices: Talking about Life and Work in the UK Since 1945*, London and New York: Bloomsbury Academic.

McGarr, P. (2016) 'Striking debates', *International Socialism*, Winter, 149: 101–24. http://isj.org.uk/striking-debates/ (last accessed 13 February 2021).

McGettigan, A. (2019) 'Fiscal illusions', *London Review of Books*, 41(17): 36–8.

McIlroy, J. (2014) 'Marxism and the trade unions: the bureaucracy versus the rank and file debate revisited', *Critique: Journal of Socialist Theory*, 42(4): 497–526.

McIlroy, J. and Daniels, G. (2010a) 'A brief history of British trade unions and neoliberalism in the age of New Labour', in Daniels, G. and McIlroy, J. (eds), *Trade Unions in a Neoliberal World*, Abingdon, Oxfordshire: Routledge: 63–97.

McIlroy, J. and Daniels, G. (2010b) 'An anatomy of British trade unions since 1997: organization, structure and factionalism', in G. Daniels and J. McIlroy (eds), *Trade Unions in a Neoliberal World*, Abingdon: Routledge: 123–64.

Migration Advisory Committee (2018) *EEA Migration in the UK: Final Report*, September. tinyurl.com/yfqdosgg (last accessed 7 February 2021).

Migration Observatory (2019) 'Briefing: migrants in the UK labour market – an overview', Oxford, University of Oxford, 19 July. tinyurl.com/1bszhv3l (last accessed 7 February 2021).

Migration Observatory (2021) 'Migrants in the UK labour market: an overview'. tinyurl.com/47z2tql7 (last accessed 7 February 2021).

Milne, S. (2014) 'During the miner's strike, Thatcher's secret state was the real enemy within', *The Guardian*, 3 October. tinyurl.com/1xwfeftp (last accessed 7 February 2021).

Mindfulness Initiative, The (2015) 'Report: Mindfulness All-Party Parliamentary Group (MAPPG)', October. www.themindfulnessinitiative.org/mindfulness-all-party-parliamentary-group (last accessed 13 February 2021).

Mirowski, P. and Plehwe, D. (2009) *The Road from Mont Pelerin: The Making of the Neoliberal Thought Collective*, London: Harvard University Press.

Moore, S. and Hayes, L. J. B. (2017) 'Taking worker productivity to a new level? Electronic monitoring in homecare – the (re)production of unpaid labour', *New Technology, Work and Employment*, 32(2): 101–14.

Morgan, R. (2016) 'A private sector perspective on the strikes debate', *International Socialism*, Winter, 149: 125–42.

Moss, J. (2015) '"We didn't realise how brave we were at the time": the 1968 Ford sewing machinists' strike in public and personal memory', *Oral History*, 43(1): 40–51.

Moth, R. (2020) '"The business end": neoliberal policy reforms and biomedical residualism in frontline community mental health practice in England', *Competition and Change*, (24)2: 133–53.

Munbodh, E. and Reid, B. (2018) 'Sport Direct UK headquarters in Derbyshire sold for £120m', *Nottingham Post*, 28 May. tinyurl.com/2zhmhbkt (last accessed 7 February 2021).

National Audit Office (H.M. Treasury) (2018) PF1 and PF2, report by the comptroller and auditor general, ordered by the House of Commons. www.nao.org.uk/wp-content/uploads/2018/01/PFI-and-PF2.pdf (last accessed 13 February 2021).

Nuffield Trust (2019) 'The NHS is the world's fifth largest employer'. tinyurl.com/lo9ptpmg (last accessed 8 February 2021).

O'Brien, M. (2018) 'What has happened to the British labour movement and what does it mean for the left in the unions?', *International Socialism*, 157: 149–73.

O'Connor, S. (2018) 'Dark factories: labour exploitation in Britain's garment industry', *Financial Times*, 17 May. tinyurl.com/e42elfve (last accessed 8 February 2021).

O'Connor, M. and Portes, J. (2021) 'Estimating the UK population during the pandemic', Economic Statistics Centre of Excellence. tinyurl.com/ukwp6q52 (last accessed 2 February 2021).

ONS (Office for National Statistics) (2018) 'Trends in Self Employment: Conclusion'. tinyurl.com/3pzq77t5 (last accessed 8 February 2021).

ONS (Office for National Statistics) (2019) 'Female employment rate (aged 16 to 64 seasonally adjusted)'. tinyurl.com/yoljf92y (last accessed 11 February 2021).

ONS (Office for National Statistics) (2020a) 'All data related to workplace disputes and working conditions'. http://ons.gov.uk/employmentandlabourmarket/peopleinwork/workplacedisputesandworkingconditions/datalist (last accessed 13 February).

ONS (Office of National Statistics) (2020b) 'Emp 13: employment by industry (last release of figures 10 November)'. tinyurl.com/yd5nj4vd (last accessed 8 February 2021).

Osborne, H. and Barr, C. (2018) 'Revealed: the developers who are cashing in on privatisation of student housing', *The Guardian*, 27 May. tinyurl.com/aptz24dn (last accessed 8 February 2021).

Parliamentary Select Committee (2019) 'Universal credit and "survival sex"'. tinyurl.com/1qdoka3z (13 February 2021).

Peck, J. (2010) *Constructions of Neoliberal Reason*, Oxford: Oxford University Press.

Pettersen, L. (2018) 'Why artificial intelligence will not outsmart complex knowledge work', *Work, Employment and Society*, 33(6): 1058–67.

Philipson, A. (2013) 'Labour made "spectacular mistake" on immigration admits Jack Straw', *The Telegraph*, 12 November. tinyurl.com/2mt8ckg7 (last accessed 7 February 2021).

Pollert, A. (1981) *Girls, Wives, Factory Lives*, London and Basingstoke: Macmillan.

Prasad, Y. (2017) 'Here to stay, here to fight: how Asians transformed the British working class', *International Socialism*, Winter, 153: 69–94.

Quinn, B. (2019a) 'Payroll "upgrade" means thousands of workers face week without pay', *The Guardian*, 19 April. tinyurl.com/1jsjhc4l (last accessed 7 February 2021).

Quinn, B. (2019b) '"Unlawful and vicious": union organiser sacked by games company', *The Guardian*, 3 October. tinyurl.com/vbu7t5aw (last accessed 13 February 2021).

Radin, B. (1966) 'Coloured workers and British trade unions', *Race*, 8(2): 157–72.

Ramanan, C. (2017) 'The video game industry has a diversity problem – but it can be fixed', *The Guardian*, 15 March. tinyurl.com/ln302h8u (last accessed 13 February 2021).

Ramdin, R. (2017 [originally published 1987]) *The Making of the Black Working Class in Britain*, London: Verso.

Randle, K. (1996) 'The white-coated worker: professional autonomy in a period of change', *Work, Employment and Society*, 10(4): 737–53.

Ransom, D. (1997) 'Interview: health workers of Hillingdon', *New Internationalist*, 5 October. tinyurl.com/106mlw56 (last accessed 7 February 2021).

Raw, L. (2011) *Striking a Light: The Bryant and May Matchwomen and Their Place in History*, London: Continuum.

RCN (2019) 'Decriminalising prostitution: why should it concern nursing staff?', bulletin. tinyurl.com/3ptqnj9s (last accessed 13 February 2021).

Rhodes, C. (2019) 'The motor industry: statistics and policy', House of Commons Library, Briefing Paper 00611, 16 December. tinyurl.com/4925rna8 (last accessed 8 February 2021).

Rienzo, C. (2015) 'Migrants in the UK labour market', Migration Observatory, University of Oxford.

Robertson, M. and Clark, A. (2019) '"We were the ones really doing something about it": gender and mobilisation against factory closure', *Work, Employment and Society*, 33(2): 336–44.

Rowbotham, S. (2006) 'Cleaners organizing in Britain from the 1970s: a personal account', *Antipode*, 38(3): 608–25.

Rowe, J. W. F. (1923) *Wages in the Coal Industry*, London: P. S. King.

Ruhs, M. and Vargas-Silva, C. (2018) *Briefing: Labour Market Effects of Immigration*, Oxford: Migration Observatory, University of Oxford.

Ruttum, L. (2010) 'The Pony Express: History and Myth', New York Public Library, 2 February. tinyurl.com/1d5fcl8a (last accessed 13 February 2021).

Ryan, F. (2019) 'How austerity is forcing disabled women into sex work', *The Guardian*, 5 June. tinyurl.com/5yfduzjk (last accessed 13 February 2021).

Salt, J. and Millar, J. (2006) 'Foreign labour in the United Kingdom: current patterns and trends', *Labour Market Trends*, ONS, October. tinyurl.com/x3jb4fc8 (last accessed 12 February 2021).

Sanders, T. and Hardy, K. (2015a) 'The political economy of "lap dancing": contested careers and women's work in the stripping industry', *Work, Employment and Society*, 29(1): 119–36.

Sanders T., and Hardy K. (2015b). 'Students selling sex: marketisation, higher education and consumption', *British Journal of Sociology of Education*, 36: 747–65.

Sasse, T., Guerin, B., Nickson, S., O'Brien, M., Pope, T. and Davies, N. (2019) 'Government outsourcing: what has worked and what needs reform', *Institute for Government*, September. tinyurl.com/12jh3dr8 (last accessed 8 February 2021).

Schwab, K. (2015) 'The fourth industrial revolution: what it means and how to respond – snapshot'. tinyurl.com/n8465ecp (last accessed 8 February 2021).

Shalev, M. (1978) 'Lies, damn lies and strike statistics: the measurement of trends in industrial conflict', in C. Crouch and A. Pizzorno (eds), *The Resurgence of Class Conflict in Western Europe since 1968*, Vol. 1, London: Macmillan, 1–19.

Sherry, D. (2010) *Occupy! A Short History of Workers' Occupations*, London: Bookmarks.

Silver, B. J. (2003) *Forces of Labour: Workers' Movements and Globalization since 1870*, Cambridge: Cambridge University Press.

Skills for Care (2019) *The State of the Adult Social Care Sector and Workforce in England: Report*, September. tinyurl.com/183tl70k (last accessed 8 February 2021).

Slee, A., Nazareth, I., Freementle, N. and Horsfall, L. (2020) 'Trends in generalised anxiety disorders and symptoms in primary care: UK population-based cohort study', *The British Journal of Psychiatry*. tinyurl.com/fw4cblh8 (last accessed 8 February 2021).

Smith, A. (2008 [1776]) *An Inquiry into the Nature and Causes of the Wealth of Nations*, Oxford: Oxford University Press.

Smith, D. and Chamberlain, P. (2015) *Blacklisted: The Secret War Between Big Business and Union Activists*, London, New Internationalist.

Smith, J. (2007) 'Shared protections, shared values: next steps on migration', speech by the home secretary, Jacqui Smith MP, given at London School of Economics and Political Science on 5 December quoted in House of Lords, Economic Affairs, First Report. tinyurl.com/yqtufk58 (last accessed on 12 February 2021).

Smithers, R. (2006) 'Exams boycott suspended after lecturers agree deal', *The Guardian*, 7 June.

Socialist Worker (2020) 'Shrewsbury 24 appeal decision shows "you have to keep fighting" – Ricky Tomlinson', No. 2695, 5 March. tinyurl.com/1lz26l6z (last accessed 8 February 2021).

Stacey, M. (2018) 'How to be a resilient doctor: skills to maximise your anti-fragility', *British Journal of Hospital Medicine*, 79(12): 704–7.

Standing, G. (2011) *The Precariat: The New Dangerous Class*, London and New York: Bloomsbury Academic.

Statista Research Department (2020) 'Amazon UK services number of employees from 2009 to 2018'. tinyurl.com/lhxvz2b1 (last accessed 8 February 2021).

Statt, N. (2018) 'Games developers look to unions to fix the exploitative workplace culture', *The Verge*, 23 March. tinyurl.com/outowt3f (last accessed 13 February 2021).

Stern, S. (2017) 'Ryanair has recognised trade unions: hell must have frozen over', *The Guardian*, 18 May. tinyurl.com/196n564w (last accessed 8 February 2021).

Stubbs, M. (2014) 'Divide and conquer: a forensic analysis of the 1894–85 cabinet papers in relation to the miners' strike', National Union of Mineworkers, May. tinyurl.com/vf14dtrn (last accessed on 8 February 2021).

Taylor, H. (2019) 'Gender pay gap widens in the UK games', *Games Industry.biz*, 12 April. tinyurl.com/274shjlm_(last accessed 13 February 2021).

Taylor, M., Marsh, G., Nicol, D. and Broadbent, P. (2017) *Good Work: The Taylor Review of Modern Working Practices*, Department for Business, Energy and Industrial Strategy. tinyurl.com/c657csat (last accessed 12 February 2021).

Taylor, P. and Moore, S. (2019) *Cabin Crew Conflict*, London: Pluto Press.

Taylor Review of Modern Working Practices, The (2017) 'Good work: independent report', Department for Business, Energy and Industrial Strategy, July, London. tinyurl.com/ctkbjh65 (last accessed on 8 February 2021).

The Angry Silence (1960) Film, directed by Guy Green, produced by Richard Attenborough, Bryan Forbes and Jack Rix, Beaver Films, United Kingdom.

The Rank and File (1971) Film, directed by Ken Loach, written by Jim Allen, shown on 20 May on BBC 1, Play for Today series. Included in the *Ken Loach at the BBC* box set (2011), United Kingdom.

Thomas, M. (2020) 'The Battle in the Workplace', *International Socialism*, 167, Summer: 35–56.

Torr, D. (1956) *Tom Mann and His Times*, Vol. 1, London: Lawrence & Wishart.

Touraine, A. (1971) *The Post-industrial Society: Tomorrow's Social History – Classes, Conflicts and Culture in the Programmed Society*, New York: Random House.

Trussell Trust, The (2020) 'Summary of findings on the impact of the Covid-19 Crisis on food banks', June. tinyurl.com/tpbqqdo3 (last accessed 8 February 2021).

Turing, A. (1950) 'Computing machinery and intelligence', *Mind*, 49: 433–60.

TUC (Trades Union Congress) (2015) *The Impact on Women of Recession and Austerity*, London: TUC. tinyurl.com/2pcfxctp (last accessed 7 February 2021).

TUC (Trades Union Congress) (2019) 'UK's gig economy workforce has doubled since 2016, TUC and FEPS-backed research shows', TUC Briefing. tinyurl.com/2lw39gfl (last accessed 8 February 2021).

UCU (2019a) 'Counting the costs of casualisation in higher education: key findings', June. tinyurl.com/4b9v26xe (last accessed 13 February 2021).

UCU (2019b) 'Sex workers and education briefing', February. tinyurl.com/1fbms9xl (last accessed 13 February 2021).

UKIE (Association for Interactive Entertainment) (2018) 'UK video games fact sheet'. tinyurl.com/1amowmdc (last accessed 13 February, 2021).

Unison (2020) 'Covid-19: black and female workers on the front line'. tinyurl.com/4uzy4rvf (last accessed 7 February 2021).

Universities UK (2019) 'Higher education in numbers'. tinyurl.com/6qfw55bk (last accessed 13 February 2021).

University of Leeds UCU (2018) 'Strike up your life! #StrikeForUSS'. https://www.youtube.com/watch?v=yvQsx-hPDeo (last accessed 13 February 2021).

Upchurch, M. (2014) 'The internet, social media and the workplace', *International Socialism*, 141: 119–38. tinyurl.com/2ls4766p (last accessed 8 February 2021).

Valentine, R. (2018) 'Eugen Systems fires workers involved in pay dispute', *Games Industry.biz*, 22 December. tinyurl.com/1vpcxs36 (last accessed 13 February 2021).

Vernon, J. (2018) 'The making of the neoliberal university in Britain', *Critical Historical Studies*, 5(2): 267–80.

Wajcman, J. (2017) 'Automation: is it really different this time? Review essay', *The British Journal of Sociology*, 68(1): 119–27. http://eprints.lse.ac.uk/69811/ (last accessed 8 February 2021).

Walker, O. (2018) 'Shared parental leave suffers inauspicious start', *Financial Times*, 8 August. tinyurl.com/2nlyk846 (last accessed 7 February 2021).

Ward, M. (2020) 'UK trade, 1948–2019', Briefing Paper CBP 8261, House of Commons Library, 10 December. tinyurl.com/7fvnsr15 (last accessed 8 February 2021).

Watson, L. (2019) '#Makeallwomensafe: a campaign changing the terrain for sex workers', *Equal Times*. tinyurl.com/1kus2seh (last accessed 13 February 2021).

Wearn, R. (2018) 'Online shopping drives the demand for warehousing space', BBC News, 27 August. www.bbc.co.uk/news/business-45210148 (last accessed 8 February 2021).

Webb, S. and Webb, B. (1920) *The History of Trade Unionism*, London and New York: Longmans, Green and Co.

Webber, A. (2020) 'Thousands of equal pay claims still received by tribunals each year', *Personnel Today*, 27 May. tinyurl.com/12si90uo (last accessed 7 February 2021).

Webber, J. and Lopez, A. (2013) 'Organising sex workers in the UK: what's in it for trade unions?', *Centre for Employment Studies Research*, University of the West of England. tinyurl.com/lfp20js1 (last accessed 13 February 2021).

Whiteside, N. (2017) 'Flexible employment and casual labour: a historical perspective on labour market policy', Policy Papers, 27 June, Warwick Institute for Employment Research, University of Warwick, United Kingdom. tinyurl.com/4jullk9q (last accessed 8 February 2021).

Williams, G. A. (1978) *The Merthyr Rising*, London: Croom Helm.

Wills, J. (2008) 'Making class politics possible: organising contract cleaners in London', *International Journal of Urban and Regional Research*, 32(2): 305–23.

Wilson, H. (1964) *The Relevance of British Socialism*, London: Weidenfeld and Nicolson.

Wolfe-Robinson, M. (2019a) 'Union stages final protest over "horrific" Amazon work practices', *The Guardian*, 22 July. tinyurl.com/1liszwpk (last accessed 7 February 2021).

Wolfe-Robinson, M. (2019b) 'Sheffield strip club keeps its licence despite opposition by feminist coalition', *The Guardian*, 17 September. tinyurl.com/1mmj5ezv (last accessed 13 February 2021).

Woodcock, J. (2017) *Working the Phones*, London: Pluto Press.

Workers Wild West (2018) https://angryworkersworld.files.wordpress.com/2018/04/workers_wild_west_7_proof5.pdf (last accessed 13 April 2021).

World Trade Organization (2019) 'World Trade Organisation statistical review'. tinyurl.com/5frf5sul (last accessed 8 February 2021).

World Bank (2019) *World Development Report 2019: The Changing Nature of Work*, Washington, DC: World Bank.

Wright Mills, C. (1948) *The New Men of Power: America's Labor Leaders*, New York: Harcourt Brace.

Index

note: *ill* refers to an illustration; *n* to a note; *t* to a table

Thanks to our Patreon Subscribers:

Lia Lilith de Oliveira
Andrew Perry

Who have shown generosity and comradeship in support of our publishing.

The Pluto Press Newsletter

Hello friend of Pluto!

Want to stay on top of the best radical books we publish?

Then sign up to be the first to hear about our new books, as well as special events, podcasts and videos.

You'll also get 50% off your first order with us when you sign up.

Come and join us!

Go to bit.ly/PlutoNewsletter